I0002889

DESIGNING PRODUCTS FOR EVOLVING DIGITAL USERS

STUDY UX BEHAVIOR PATTERNS, ONLINE COMMUNITIES, AND FUTURE DIGITAL TRENDS

Anastasia Utesheva

Apress®

Designing Products for Evolving Digital Users: Study UX Behavior Patterns, Online Communities, and Future Digital Trends

Anastasia Utesheva
Mullumbimby, Australia

ISBN-13 (pbk): 978-1-4842-6378-5   ISBN-13 (electronic): 978-1-4842-6379-2
https://doi.org/10.1007/978-1-4842-6379-2

Copyright © 2020 by Anastasia Utesheva

This work is subject to copyright. All rights are reserved by the Publisher, whether the whole or part of the material is concerned, specifically the rights of translation, reprinting, reuse of illustrations, recitation, broadcasting, reproduction on microfilms or in any other physical way, and transmission or information storage and retrieval, electronic adaptation, computer software, or by similar or dissimilar methodology now known or hereafter developed.

Trademarked names, logos, and images may appear in this book. Rather than use a trademark symbol with every occurrence of a trademarked name, logo, or image we use the names, logos, and images only in an editorial fashion and to the benefit of the trademark owner, with no intention of infringement of the trademark.

The use in this publication of trade names, trademarks, service marks, and similar terms, even if they are not identified as such, is not to be taken as an expression of opinion as to whether or not they are subject to proprietary rights.

While the advice and information in this book are believed to be true and accurate at the date of publication, neither the authors nor the editors nor the publisher can accept any legal responsibility for any errors or omissions that may be made. The publisher makes no warranty, express or implied, with respect to the material contained herein.

Managing Director, Apress Media LLC: Welmoed Spahr
Acquisitions Editor: Shiva Ramachandran
Development Editor: Rita Fernando
Coordinating Editor: Rita Fernando

Cover designed by eStudioCalamar

Distributed to the book trade worldwide by Springer Science+Business Media New York, 1 New York Plaza, New York, NY 100043. Phone 1-800-SPRINGER, fax (201) 348-4505, e-mail orders-ny@springer-sbm.com, or visit www.springeronline.com. Apress Media, LLC is a California LLC and the sole member (owner) is Springer Science + Business Media Finance Inc (SSBM Finance Inc). SSBM Finance Inc is a **Delaware** corporation.

For information on translations, please e-mail booktranslations@springernature.com; for reprint, paperback, or audio rights, please e-mail bookpermissions@springernature.com.

Apress titles may be purchased in bulk for academic, corporate, or promotional use. eBook versions and licenses are also available for most titles. For more information, reference our Print and eBook Bulk Sales web page at http://www.apress.com/bulk-sales.

Any source code or other supplementary material referenced by the author in this book is available to readers on GitHub via the book's product page, located at www.apress.com/9781484263785. For more detailed information, please visit http://www.apress.com/source-code.

Printed on acid-free paper

For all that is alive.
For Dennis.

Contents

Contents

About the Author

Anastasia Utesheva has dedicated her career to improving quality of life through strategic design. Her experience working across academia, government, and private sector organizations has shaped her pragmatic approach to sustainable systemic change. Anastasia believes that the coevolution of humans and technology is core to transforming legacy ways of being and creating truly beautiful and sustainable ways of life. Anastasia specializes in shifting thought systems to empower creators to (re)design through empathy and core value alignment.

Acknowledgments

I sincerely thank Shiva Ramachandran and Rita Fernando for their guidance and encouragement in the publication of this work. I deeply appreciate your perspectives, enthusiasm, and kind help in expressing the most complex of ideas.

I also want to thank the staff of Apress for the opportunity to embark on this project and for their assistance in the publication.

Thank you to my friends and family for making the journey that much better. For your love, your patience, and your ability to make me see everything anew.

Finally, I wish to express my gratitude to everyone from whom I have had the honor of learning. Without you, the ideas would not be of the form they are today.

Introduction: Design for Life

"For everything to stay the same, everything must change."

—Unknown

We live in a world of exponential change. The ripples of change redefined everything we thought we knew, including how we think about ourselves and our role in our global social reality.

Over the past century, digital has become a core element in shaping every part of our reality. What we felt most keenly was that digital has allowed us to collect and carefully curate information about ourselves. We can now control others' perception of us through the social media posts we share. We can look for any information available on a prospective client before ever entering the meeting by stalking them using Google. We can view the daily life of our favorite vlogger on YouTube. We now compare habits and values from all cultures, systems of thought, bodies of literature, emerging technologies, and countless voices in digital communities.

Recently, something changed in our world, but we hardly noticed. We became a composite entity, part human, part digital. Unlike the dystopia that we apprehensively envisioned in the early days of digital, this reality is one of coevolution. A careful balance of beauty and horror, as we see more and more of ourselves reflected back at us through our digital extensions. As our world transitioned from analog to digital-first, we cleverly created a hyperrealism where digital became more important to us than the analog reality that birthed it. We started to live primarily in the reality enabled by digital. Many digital spaces, many as apps on your phone, all intertwine and merge to become one cohesive reality through their use. Instantiated through the pronoun "I" and changing with each moment of interaction with yet another space that extends our material existence, the two halves that once were (i.e., the analog and digital identities) rapidly became one.

Digital media has enabled us to see, share, empathize, collaborate, and create in significantly new ways. It has allowed for us to trace occurrences and global patterns with remarkable accuracy. We gained a way to see our Earth from its orbit; a protest in the next town over; a sports game live we couldn't get tickets to. A way to see the house we grew up in without having to leave our new living room, all the while shopping the latest New York fashion. Digital allowed us to transcend the temporal and spatial limits of our bodies, to see and enact change all over the world no matter where we physically inhabit. It has guided us on how to behave and how to interpret the world in broader, more nuanced, and more meaningful ways.

While being alone with our glowing screens, modern humans are constantly connected to a buzzing global network of people, ideas, and opportunities previously beyond our reach. The local has been made global, the global accessible to any locale. To enable connection, purely digital, immaterial, spaces came into being. By creating and adopting digital, we created spaces that have not existed before. Whole digital ecosystems, some of which became an extension of the literal and some of which remained separate in their own right (outside of our engagement with them). They are everywhere and nowhere. We create instances of them when we open our browser or app, and those instances co-create our reality. In this manner, beyond ideas, digital products have shifted our very habits, becoming integrated into our lives to the point of inseparability.

We realized quickly that at the root of most change is *information*. Transparency or obscurity. Abundance or scarcity. Due to its unprecedented convenience, accessibility, and trustworthiness, our go-to for information has shifted from people to technology. Before, we relied on humans. Now, our questioning starts with: *Is there an app for that? What does Google say?* Greater access to information has fundamentally transformed our relationship with our world and our place in it. We are still acclimatizing to the changes.

As digital quickly became a literal extension of us, our sense of self and identity adjusted to the changes we experienced. We suddenly needed to position ourselves in relation to all others on the global stage. We had to adjust our perspectives of quality and value, of the purpose and impact of what we create. Prior to digital, who we thought we were and our values were largely dependent on the limited information we could access that helped us contextualize and make sense of our lives. With almost unlimited access to information, digital has fundamentally transformed this foundation.

Identity narratives (i.e., stories we tell ourselves about who we are, and what we do and don't do) shifted to encompass new tonalities of our being: *Am I an Apple or Android kind of person? What's my position on mobile phones at the dinner table? Who do I know on social media? What kind of personal brand is this Instagram post cultivating?* Though identity narratives are not a new phenomenon, their scope, complexity, and rate of change have transformed dramatically in the last century. Identity narratives have helped govern behavior, assist in feeling a sense of purpose and value, and identify with those we consider "our tribe." Who we are has become more driven by the reflection we see of ourselves in the digital than those that surround us in our geographical locale. We now construct and design ourselves based on countless role models and archetypes that we have access to.

In one way, we freed ourselves from having to accept linear evolutionary trajectories and disciplines of thought that characterized our past. At the same time, we have overloaded ourselves with choice while forcing ourselves to reexamine the very nature of our being. This is a good thing. A profound state of empowerment that we, as a species, have not had a chance to experience before on a global scale.

As digital designers, we are keenly aware of these constant changes. We comprehend that the users we design for are part of a global myriad of interconnected, often symbiotic, networks, some of which do not exist outside the digital realm they constitute yet are more important to our users than anything else. The only interface to these realities is through the digital products and/or services that allow us to interact and enact change within that space (and beyond).

As digital designers, we have a responsibility to our users to create the optimal experience for them. An experience that enhances their lives, brings about positive social change, and enables them to go beyond what is possible. As our users change, we need to preempt their changes and allow them to transition to a new state of being through positive experiences. We act as enablers of their dreams, their aspirations, their needs and wants, their feeling of safety as they navigate the external, full of new challenges and complexities. Our goal is to enhance their lives, make them easier, improve quality, and add to elusive feelings of happiness they are searching for.

In creating digital products and services, digital designers become the leaders of the changes that constitute our modern world and our identity. That is a lot of responsibility.

So how do we do it? By designing for life.

Our Digitally Distributed World

Life at the turn of the 21st century marked a new era. An era of true digital extension of ourselves. Unlike ever before, we developed a symbiotic relationship between us and the tools we created. Digital has become a part of who we are, an enabler of what we do, and a constant interface between us and "other."

Digital has become inseparable from our identity. It became invisible, something that is so common that it simply *is*, hidden in plain sight even for those of us working to create within its boundaries. Since the power of language in transforming our entire being, it is hard to point to something more profound than digital technology. This is something that comes as no surprise to those looking at the past through the lens of exponential evolution and refinement of information. The more sophisticated our information ecosystem, the more advanced we become.

Digital has rewired us, unified culture, broadened horizons, and made time and space irrelevant in participating in global affairs. New patterns of being have emerged. With Google, we can *find, see, compare, watch, listen,* and *learn.* With Facebook, we can *cultivate, discover, connect with, see, watch, self-express, record, curate, discuss,* and *dream.* With Amazon, we can *reach, acquire, create,* and *save.* With eBay, we can *downsize, upsize, search for, hunt, get,* and *sell.* With SecondLife, we can *become, experiment, explore, discover, test, create, destroy, play,* and *connect.* With Reddit, we can *argue, share, delight, learn, connect with,* and *empathize.* With Netflix, we can *escape, relax, immerse, suspend, discover,* and *feel.*

We can do more than ever before, everywhere, all the time. We can cycle through different modes of being, no matter how different, in seconds. We can see ourselves reflected to us through each pattern of interaction. As our knowledge of ourselves grows, so do we. Nothing encourages self-reflection and reflexivity more than transparency over our own choices. We are now free in ways we have not been before. And we have gradually become more mature, empathetic, aware, conscious, and caring. Paradoxically, at the same time, we have become more disillusioned, disassociated, overloaded, over-worked, and burnt out.

We have developed incredibly high standards for ourselves and our expectations of others. We know what the "cutting edge" standard is at any time, and we are not willing to compromise on it without a good reason. We, as users, have developed increasingly high expectations of what a digital product or service can offer us. Due to the amount of choice we have, anything that creates a hindrance to our progress (in performance, functionality, or UX) becomes instantly replaced by something that is more aligned with our preferences.

Successful products and services became leaders of global change in standards and user expectations. Any one successful innovation generates waves of adopters, who follow the new standard to make sure that they do not become obsolete, as some inevitably do. Global waves of innovation merge in the melting pot of digital, and create new patterns of life, which can be both positive and negative from a holistic perspective.

The changes that have resulted in the positive gains over the last century also created a host of pain points that we are still working to design out of the system. Technological progress has significantly improved quality of life for a lot of us, while at the same time created waves of ecological instability and pandemic existential crises in even the most highly developed societies. As digital designers, our power is in the level of impact that we can achieve through the products and services that we create. We can, in theory, use technology to solve the most complex "wicked" problems that we face within our lifetime. We are aware of the issues that are still preventing groups from achieving a state of symbiotic sustainability with the broader ecosystem. We are also aware of the responsibility we have in leading the transformation.

Just as we became more empathetic as a society, we became more empathetic as designers. We now learn from our users, rather than dictating what their experience should and should not be. We have more ways to see what users do in the digital spaces we create, and interact with them as they interact with the product and/or service created. The more information comes to light, the more we are able to make an informed decision for the benefit of our users and their purpose. The choices we make, as digital designers, are more complex than ever before.

The more we are able to see the similarities and differences, the more we are able to comprehend that what separates us is as meaningless as a candy wrapper, but what unites us is so fundamental that we can all agree to its indispensable relevance to our existence. We recognize that to be human is to be humane, to think beyond our short-term interests, and to design our role in our collective evolution.

Digital Design

Welcome to the exciting world of digital design. If you are a veteran forging its path, or someone with a keen interest, welcome to the most exciting design project of our time: digital. It started in the shadows, out of the main bustle of business activity. Then it became a part of every office, every meeting, every mass production operation, every home. Digital has proliferated and become embedded in everyday life to the point of inseparability with those that use it.

At all points in its evolution, digital had one purpose: to enhance the lives of those who interact with it. Though originally digital was thought of as a mechanistic "thing," it soon became something that is an undeniable extension of us. Our phone and laptop are now something without which we cannot function in modern society. There are still overlaps with prior waves of society where digital technology did not play such a critical role. The mindset of the past is one of fear of change, while the modern fear is one of stagnation. We no longer want to be anything other than changing rapidly, to keep up, to become something better. We cannot stand to be kept out of the new wave of progress. We don't want to become irrelevant, or worse, unfashionable or out of touch.

We are now designing digital for evolving users. We are creating something mostly intangible: an experience. As it emerges, it has effects that we deliberately designed into the system, and others completely unpredicted. Something as simple as a "Like" button and where it is placed on a page (hierarchy of importance) can have unintended consequences of shaping user behavior, such as to focus on posting content that aims solely to boost its own popularity (and also its dominance in the idea pool).

As designers, we make countless design decisions in every project that have the same dilemma. We can predict the impact of features on user behavior only so far. We are still very limited in predicting how the product or service will truly impact someone's life, how it may shape their perspectives and habits. Or the more personal questions: *How might interacting with this product or service shape how users will feel about themselves? How might the product/service affect the formation of users' identity?*

Identity in the digital space is complex. Before we could represent ourselves online, we primarily had a static view of identity. Who we were was closely linked to our material bodies. Identity came from biological relations, and those of similar agents in the society that we constituted. Then digital allowed us to be 100 people and 1 at the same time. There are no limits of the number of identities, profiles, and avatars that you can have in the digital realm. We gained digital identities and started to question the "other" identities that suddenly became apparent as, perhaps, not being a 100% complete representation of ourselves. We realized we are far more nuanced and complex than we originally thought. We realized that who we are is more a product of our choices than the static attributes that defined us before (e.g., gender, age, wealth, status, etc.).

Who we were online bled into who we were offline. The two are now the same. Digital design has enabled this to happen.

Identity in Digital Design

Identity is a narrative and/or a set of attributes we use to define ourselves. Identity is the filter that interprets our sense of self and enhances it as it sees fit in the memories that it curates. It is something that we develop over time, since birth, and is the outcome of our comprehension of our world and our experience in it.

The most useful way to think about identity in digital design is as an algorithm that works within certain sets of parameters. Certain inputs are interpreted through existing schemata we have cultivated over time and recognize as "I". We can predict our own actions based on what we know of ourselves. We can, to a degree, predict how we will be perceived by others. We now know what drives us, and what we can and cannot do based on our personal ethics. We know what values we uphold and what we are willing to compromise at what price. We realize that who we are is a product of what we choose, rather than what we were born into.

No two humans perceive the world the same way. How identity is constructed depends on what someone identifies with/as, and how that formed throughout their life. Someone may identify as their emotions, another as their thoughts. Another may identify as the role they play in society, their gender, their social status, or their achievements. Some may identify themselves as a part of a group, like a family or business. Others may identify with their personal choices, their lifestyle choices, or whom they choose to associate with. Some may identify themselves as their looks or their biology. Some identities resemble complex stories where the protagonist "I" acts differently in each context, yet maintains their coherent self through upholding of values and ethics. Others are based on social media profiles, abstract ideas, trends, or political movements. And someone else may identify as the persona(s) they are online, more than anything else in their reality. Each identity is constructed out of a complex set of parameters and interpretation algorithms.

To design for identity we need to examine *how* and, more importantly, *why* users make choices and how those choices reflect their identity. We need to comprehend how to orchestrate the interplay of action and content into a seamless frictionless experience that delights and improves the quality of life of those partaking in it.

As digital designers, we have learned to pay close attention to different paradigms that emerge through the affordances and constraints of the digital products and services we create. We began to realize that what users can and cannot do creates certain pathways for them to follow, like water flowing through textures of terrain in a valley. We realized that how we designed the product or service to behave had direct impact on the behaviors of users. We realized that what we rewarded and encouraged in the digital space skewed community evolution around optimizing those characteristics. We realized that the identity of someone has more of an impact on what they say they do than what they may actually do.

To design for identity through affordances and constraints of digital requires us to comprehend where our users are and where they aspire to go. Then, to create truly meaningful digital products and/or services.

How might we design for identity? How might we out innovate competitors? How might we create phenomenal products and services?

Design for life.

How to Use This Book

This book provides an exploration of how to design digital for evolving users. Depending on where you are in your journey, this book can be read in a number of ways.

The first way is to read the chapters in sequential order. This reading provides a high-level overview of trends in technological evolution, followed by a discussion of the evolution of human identity. Then we delve into how to design for identity, how to get to know users through Design Thinking, and how to create meaningful and purposeful products and services. The book concludes with an overview of trends that we currently experience and where we may go next.

The second way to read this book is by topic. The chapters provide stand-alone discussions, so reading by interest will not undermine the value of the book.

The third way to read this book is to read the chapters in reverse order. This way is most suitable for practitioners who just want to get to the applicable methods, tips, and insights, and get started designing. The last three chapters provide practical insights, while the first three provide the context and conceptual exploration of the phenomena discussed.

I hope you enjoy this book and take away something practical to apply to your digital design practice.

Digital Evolution

Coevolution of Humans and Technology

We are a product of what we create, as much as what we create is a product of us. The last two centuries have brought into sharp focus a curious relationship between humans, cultures, ways of thinking, and the technology that we create, appropriate, and normalize into our lives.

Digital technology is evolving at such a phenomenal rate that shifts in global social dynamics can now be seen in less than 12 months. There is increasing public recognition of the exponential nature of digital progress and innovation. One of the most obvious examples is the reflection of Moore's Law in technological progress through the observation that, since the 1960s, the number of transistors in a dense integrated circuit has doubled roughly every 2 years while the costs of computers has halved. Along with the changes in price-performance of computation (i.e., as cost per unit of production decrease and number of computations per dollar increase), more and more powerful digital technology is becoming globally available and adopted. Our mobile phones are more powerful now than supercomputers were in the mid-20th century. They are also far cheaper and more accessible. The social changes due to this trend of technological advancement are vast and often overlooked as the use of digital technology becomes a more and more constitutive part of our everyday lives.

© Anastasia Utesheva 2020
A. Utesheva, *Designing Products for Evolving Digital Users*,
https://doi.org/10.1007/978-1-4842-6379-2_1

Now we carry around more computational power than we once thought possible, and what we found ourselves doing with it has been equally as surprising. We are now able to reach, see, create, and influence more than previously possible in history. Digital media reflects our increased connectivity and our adaptation to new forms of being. We now live in a time where information is more obviously significant in our lives than ever before. It is what we crave, what we go to extremes to get, what we base our sense of self and identity on, and what we build our habits around. This increased desire for connectivity has enabled us to explore ourselves more fully than ever before, and our exploration of ourselves has aligned our desire for self-expression with the desire to design our technology to suit our needs (rather than to constrain us).

To comprehend the *how* and *why* of the social trends we embody and design for, we need to look at the changing nature of the relationship between digital technology and humans. Such changes are not limited to digital innovation or the increasing complexity of information. Global social dynamics are becoming more complex as a result of technological progress. We can now *see* and *feel* change within one generation. These are changes that go beyond the superficial and to the core of who we are. Lingering assumptions of gradual progress or long-term stability are becoming increasingly difficult to maintain. We constantly witness the emergence of social, technological, and cognitive paradigm shifts that reflect the exponential rate of progress and, as is often the case, its misinformation by-products that we have become wrapped up in. We now have more information to access, more to comprehend, more to create than ever before. And we cannot do this without technology. We cannot evolve in the way we want unless we comprehend the role that technology plays in our latest evolutionary step.

How does this all relate to designing products?

Types of Information

To design for a user is to design for a complex, almost foreign perspective of *someone else* or "other." In order to do so with any success, we, as designers, need to look beyond the superficial and see this individual as part of a complex system: a system of interrelated and constantly evolving information types.

To help untangle the different types of information that constitute our new technologically enhanced selves, we need to look at four different, yet interrelated elements of our complex reality: biology, technology, culture, and cognition. These four types of information constituting our being are important to keep in mind in digital product and/or service design, as we are not static beings that do not change over time. We are in fact the opposite, and to forget that is to design products and/or services that simply do not resonate or add much value to users' lives (or for long).

How might we design for purposeful and meaningful change and, most importantly, assume our full responsibility in creating and driving these changes?

We first need to immerse ourselves in the patterns of information we see evolving and deeply comprehend how and why these changes occur.

Let's look at an example to illustrate this point. Consider the following technology interaction: a 37-year-old woman from New Zealand is playing a game on her phone. If we just look at the surface, we reduce our analysis of the interaction to what we can literally observe in front of us. Our view here: it's simply a person who is interacting with a product. Adopting this reductionist view, we do not take into account that the user is female, of a certain age and nationality. We reduce our perspective of the interaction from someone who has a rich history, intent, and purpose to someone with thumbs who can interact with the phone. If we then only focus on functionalities within the game without considering who is playing and *why* it appeals to them over another game, we do disservice to product design again. At best, we know what the game does and how someone can play it, and design within these parameters of our focus. At worst, we make all the wrong assumptions when we redesign or change the game so that our design decisions make the game less appealing to this human, rather than more. If we do make this game more appealing, are we using positive reinforcement that betters the human, or are we using patterns of addiction to make it harder for the user to stop engaging with the game? Limited perspectives on evolving phenomena typically skew product design in an unhealthy and/or unethical way, and may be a significant contributor to how products become obsolete and/or become stuck on a path of incremental, linear, change.

Alternatively, we can look at the user as an evolving being on a journey that is unique to them, acknowledge their history, and strategically design a product/service to enhance their journey rather than hindering it or unintentionally skewing their behavior and/or perception. We look at the woman playing the game and ask: *why* is she doing it *at this time, in this way?* We would ask her to reflect upon her choices, help us (and likely her) gain a deeper insight into her internal decision making processes, and evaluate her feelings toward the product and its place in her life. From there we would be able to "get to" the root cause and effect of her interacting with this product, where it fits in her life, and how she feels about it. Unravelling core details about the lives of those we design for is fundamental, as the more we know about them and how they came to be, the more we are able to design products around their core drivers, the core themes that govern who they are and form their sense of self, and the identity that they develop to comprehend and operate in the world. More importantly, we can begin to answer the question: how did this person come to be who they are (and not someone else)? How might this product/service fit into their lives? How might we design a better product and/or service that fits their specific needs and wants?

To consider these crucial details and use them effectively in design, the exploration of four different evolutionary streams of biology, technology, culture, and cognition is a useful starting point. Biology refers to the physical form of living beings. Technology refers to any innovation that extends the "reach" of life beyond the baseline of biology. Culture refers to the information set pertaining to social interactions among members of the same and/or similar species, which were traditionally inherited through birth into the culture and now are converging in a single culture of the "digital global." Cognition refers to the information sets and processes through which a being operates in order to comprehend and function in the world.

Biology is the most consistent of the four in its rate and method of change. Technology has been a constant companion of our species, enhancing our reach beyond what we may otherwise biologically be limited to. Technology here refers to both material (e.g., hardware, tools, machines, etc.) and immaterial (e.g., language, class structures, social norms, belief systems, abstract reason, logic, digital, etc.). The line between biology and technology is arguably growing more and more blurry over time, as the boundaries between human and digital dissolve through their ongoing entanglement (i.e., continuous coevolution). Culturally and cognitively humans have archives of historical data illustrating how and why we, as a species, have progressively been changing. The major shift in the 20th century that accelerated the observable changes has arguably been digital technology (our closest mirror to life we have been able to create completely through human effort).

Consideration of the exponentially increasing rate of change is important in untangling and comprehending the elusive causality of our reality, of life in an increasingly complex world. We can now observe the intentional way in which humans are placing less emphasis on traditions and locales, and instead connecting and evolving through digital. So much so that this is starting blur lines between the digital world and the material world that birthed it. Without a stable cultural base, we have also become free to think in ways unseen before and to explore and create new ways of thinking.

In a lot of ways, digital was the atomic bomb that converged all culture, wisdom, knowledge, information, and, most recently, data. Since the wide adoption and proliferation of this technology, we have stopped being "local" and started to form our sense of self (and our identities) through interaction on the global stage. We now have a relatively uniform pastiche of culture we can relate to no matter where we are geographically. A lot of localized traditions have died out or have become integrated into the global set. We are now united through common values, ethics, and preferences, rather than the minor details in which we enact them. What this means is that we no longer embody the disparate cultures we were born into and instead pick and choose what we want to uphold and be. Through this process, we have become more reliant on our identity (i.e., the set of attributes we use to classify ourselves), in the way we perceive ourselves and make sense of what happens.

Because identity has become so important to the modern human being, let's get a bit technical on how this new conceptualization of identity fits into our evolution. Let's assume for a moment that information is constantly evolving. There are four types of information that converge as one human being: biology, technology, culture, and cognition. In essence, the human sitting near you using a mobile phone is the convergence in the material of billions of years of this evolutionary process. As are you. Pretty incredible to think that we are what this evolutionary journey has culminated in thus far. What we do with it once we have had this realization, well, that's the true test for intelligence.

The core assumption in this perspective is that information evolves over time in different forms through different media. As its complexity increases, information transforms into new forms. As information is refined over time and media, the patterns that remain the *same* and *different* become the focal point. For example, humans have communicated with one another for centuries, yet this century is the first time that they can do so over digital text to someone that they may never meet in person. What is the same here is the need to communicate. What is different is *how* and *to whom*. The *why* may be the same or may be different. We won't know unless we explore the details of each interaction and compare the core drivers, themes, and value of the experience.

To comprehend the *why*, we need to keep in mind that the reality we live in is relational. As it is relational, we cannot assume that there is a single "truth" out there, some unchanging dataset that always was and always will be. Most of what, we as humans, have considered "truth" turned out to be nothing more than a collection of fragmented information generated from obscure and often heavily biased sources (usually an "authority" of some kind). We as a species, in the last 60 years in particular, have invested a significant amount of time and effort finding our way out of the global misinformation matrix we created. We were able to see that the reason that we used to propagate and mandate biased/linear assumptions (even to the point of threatening life) is that, historically as humans, we have been inherently biased through skewing perception of reality to benefit the few at the expense of the many. This is why, in our most recent focus on liberating information, it has not been an uncommon pattern in most cultures that what we may consider "fact" today may be considered "false" tomorrow. In the digital age, there is no longer such a thing as a "single source of truth" and its accompanying concept of "authority." We now look for "experts" (i.e., those who have spent considerable time growing and refining a knowledge base for the benefit of all), knowing that "authority" is a fabrication and has, historically, often been skewed by a few to attempt to control others.

"Truth" for a designer is not a hard unchanging fact written by one individual without explanation or explication; rather, "truth" is a measure of the *resonance* of someone with an idea, a feeling, or a concept. As symbolism and a symbolizing mind has been a core technology to fuel the evolution of culture, technology, and cognition, the concept of "truth" needs to be viewed as emerging from a specific context and the local variations resulting in it. Hence why focus on change and historicity is key to design. What we design today may not be useful tomorrow. Or how we imagine the future today may be the opposite of what tomorrow actually brings. However, if we look at patterns of change over time, we may become better equipped to adapt to change as it comes. To *design for* it, if not to fully predict it.

To learn how to design for evolving users, let's adopt compatible relational evolutionary perspectives of information. Evolutionary perspectives focus on changes among patterns and fluctuations in context that constitute those changes (rather than linear perspectives that attempt to predict the future based on the information collected in the past, typically implementing linear assumptions). Evolutionary perspectives assume constant holistic change. For this reason, evolutionary perspectives are valuable as a basis for digital design for evolving users. This chapter presents such perspectives through the framework of Universal Darwinism and provides us with a way to view how we evolve, where we came from, and how this came to be. We can then use this perspective as a basis for comprehending and designing for evolving users.

Universal Darwinism

The concept of evolution originated in observations of biological life on Earth, and the core mechanisms of which have been applied to many areas beyond the original theory of biological evolution. The concept has consistently provided valuable perspectives on phenomena, which account for relational causality over time. For example, you can study the evolution of building styles in a neighborhood, or how your chocolate chip cookie recipe changed over the years. You can look back through childhood photos and see how your body, your sense of style, and your feelings and perspectives have shaped who you are today. Evolution is the concept of change over time, such that the change enables (or constrains) adaptation to changes in the entities' internal/external context. In a sense, the process enables the survival of the *fitting* (i.e., most suitable to the broader whole).

The mechanisms and processes of evolution vary greatly among media (e.g., evolution of biology vs. evolution of the automobile), yet the overall pattern remains the same. This is why we can successfully use the meta-theoretical framework of Universal Darwinism to explore changes and patterns in all forms of media in a specific, analytically bound, context.

A meta-theoretical framework is a way to see similar patterns in different perspectives, which form the assumptions and mechanisms through which phenomena are examined and interpreted (and designed). It is a template of sorts that we can use to select different evolutionary theories and see whether or not they are compatible (i.e., whether they operate from the same set of mechanisms and/or assumptions, and can be used together). Let's start with a brief overview of the framework, focusing on its core mechanisms, and common misconceptions.

Popularized by Richard Dawkins in 1982[1], the term Universal Darwinism refers to a variety of approaches and theories that extend the original theory of evolution[2] beyond its application to the domain of biological entities on Earth. Universal Darwinism focuses on change in entities of a certain type over time and in their context. Although the mechanisms for the specific entity evolving may be different (e.g., biological change happens very differently to changes in clothing choices in a fashion magazine), the mechanisms constituting both patterns of change are similar.

Universal Darwinism posits that evolution occurs in any group of entities if there are mechanisms for introducing variation, a consistent selection process, and mechanisms for preserving and/or propagating different patterns in the "population." Universal Darwinism can be applied in any domain as long as the patterns that constitute the changes can be explained through the core mechanisms of evolution (in their most abstract form). A set of genes (i.e., biological information segments that constitute the formation of said entity) can be viewed to change through the same abstract patterns as a news article on the Web (i.e., digital information segments changing over time to shape the story)—both start off as a specific pattern at a point in time and, depending on what happens to them and other entities in that context over time, change in pattern to be most suitable to their context (survival of the fitting) or stop existing if they are no longer relevant or advantageous (extinction). Think of the evolution of a breaking news segment over time—it changes as the selected information and its presentation generates the emotional charge and the political spin of the story.

It is important to note that Universal Darwinism itself is not a theory that can be used to explain or predict low-level phenomena (e.g., survival of a company or specific behavior of an individual), but is very helpful in comprehending *changes* in trends and patterns in specific sets of entities over time and to compare changes in very different sets of entities. In essence, we look for patterns of change in hindsight (rather than attempting to make predictions) and see patterns emerge that indicate the core mechanisms at play. Upon comprehending the core mechanisms, we can design accordingly to enable, evolve, or disrupt.

Indeed, specific theories and mechanisms for biological evolution cannot adequately explain or predict technological, economic, cognitive, or cultural evolution, and vice versa. However, they can provide a useful way to view

---

[1]Dawkins, R. (1982), "The Extended Phenotype."
[2]Darwin, C. (1859), "On the Origins of the Species."

change, especially when we look at change in what we can call a "composite entity": part digital, part biology (e.g., a human user interacting with the Skype app on their phone).

In the digital product world, we are constantly having to draw boundaries between what is digital and what is not, and often come across challenges in the distinction. When a digital product goes from concept to implementation (i.e., a live website or app that humans can interact with), it literally transforms from an abstract idea in someone's head, to a set of emails, to a set of whiteboard drawings and post-it notes, to a set of requirements/user stories, to a set of designs, to code, to a prototype, to a live product. Not to mention a whole collection of supporting documentation (and social affordances and constraints) that marks its evolution across media (human/analog/digital) over the course of the project. The humans that are the intended end users use the digital product and it changes again, with their feedback and what the end users (and the product team) deem as "improvement" in subsequent releases. So what is the digital product in each of these stages of change? To meaningfully examine that, we need to look at the patterns of change in and of themselves (i.e., what remains the same and what changes over each instantiation).

To do so, we need to be a little defiant with traditional notions of linear causality (i.e., A causes B, B causes C, etc.). Rather than linear and simple cause and effect, we look at causality as relational and multiplicious. An outcome or material instantiation of a digital product at a certain point in the project (e.g., a set of UI designs) is assumed to have many influences and mechanisms for it emerging over something else that could have happened.

As such, Universal Darwinism adopts the principle of "universal causation," which implies that every event has a cause (i.e., there cannot be an event without a cause) but the cause cannot be assumed to be linear, singular, or simple to identify. In metaphor, think of a wave in an ocean, rather than a trigger on a gun. It is important to note that this perspective does not undermine the potential for novelty or imply that every event is predetermined and predictable. Quite the opposite. The focus becomes on tracking patterns of change to influences/influencers that have more "gravity" (i.e., are more prominent over time in that context and the most change can be traced to). There may be more than one (e.g., a stubborn product owner pushing their own worldview *and* budget constraints in that financial year) that skew the evolution of a digital product in a certain direction (e.g., to have a certain set of product features or look a certain way).

In the example of digital product emergence, we can see that what evolves over time is not only the product (across varying media) but also the humans involved (through their knowledge, preferences, and ability), the company or group they are a part of, and eventually the end users and the wider social dynamics the digital product constitutes. To meaningfully be able to see what works better and to design digital products that have positive social impact, we need to examine a complex and constantly evolving tapestry of changes. To make this analysis possible, these changes can be delineated for our purposes analytically into four types: biological, technological, cultural, and cognitive. Examining these four types, we need to focus on recurring patterns, especially when comparing phenomena at multiple levels of analysis and/or rates of change. Hence, the Universal Darwinism framework is helpful, as it provides a means to examine the symbiotic relationship of biology, technology, culture, and cognition through focus on similar patterns of change in different entities at different analytical levels.

Biological Evolution

At the core of some of the largest historical paradigm shifts in both scientific and social realms is the theory of biological evolution by natural selection as first proposed by Charles Darwin in *On the Origin of Species* (1859). Challenging the dominant assumptions of the time (i.e., those of fixed or constant typological "classes" in biology that accorded with an unknown divine plan), the theory presented a unified way to explain and track the observed diversity of biological life on Earth. Although the idea of evolution itself was not new (it emerged in many recorded forms, going back to the ancient Greeks), the major contribution of Darwin's perspective was the overcoming of a powerful Platonic bias against the idea of descent of seemingly unrelated species from a common ancestor. Pre-Darwin, "essences" or forms of beings were assumed to be unchanging: a being couldn't change in "essence" and new "essences" couldn't be born. A reptile could not *turn into* a bird, no more than silver into gold. This perspective was problematic for explaining and classifying the large diversity of similar biological forms and any observed changes over time in already classified species.

To help classification of species and to better comprehend how different species came to be, Darwin (1859) proposed an alternative view. In his view, differences in biological forms stem from the process of natural selection where traits that enhance survival and reproduction chances of an organism become more common in subsequent generations. Darwin spent years studying different bird species in different neighboring habitats and observed that members of the same species looked and acted differently depending on the specific habitat he found them in. He also observed that the same species' behavior changed over time, as one bird learned something new and others learned from it (e.g., changing bird song patterns). Darwin believed that the gradual process of accumulation of small changes, over enough time, could create speciation (i.e., the divergence of a genetic path to a "new" subspecies). This perspective sparked a pivotal transition in scientific thought concerning the very nature of biological change and our own historic journey of becoming human.

Human beings began to see themselves not as an entity that was "designed" by God and left unchanging over time, but rather as something that emerged through a long process of accumulated change from simpler life forms that came before it. Change that resulted from gradual adaptation to the broader changes to other life forms in our habitats. Although different, human beings began to see themselves once again as part of nature (along with all other life forms on Earth). As changes to our habitats affected our access to resources, those most capable of surviving and thriving despite (or because of) the changes multiplied in numbers, while those that were not able to do so as well decreased in numbers. Over a large enough period of time, large shifts in our form and behavior became observable, as any behaviors and/or characteristics that provided an advantage (no matter how minor) became amplified over time as they were "passed down" through subsequent generations.

Darwin also argued that multiple species can evolve from a common ancestor, each species exhibiting divergent heritable traits resulting from their unique adaptation to changes in context and their capacity to refine the advantageous traits through future generations. The same species of plant, over enough time, may look and require completely different environmental conditions to thrive, if planted in very different habitats. Darwin collected empirical evidence to illustrate how individuals within a species most able to adapt to contextual changes survive and reproduce, thus passing their adaptive advantage to subsequent generations in an iterative ongoing process (for detailed analysis and examples, see Dawkins 1976)[3]. As this perspective proved very useful in explaining the patterns that Darwin's contemporaries observed in the flora and fauna they studied, it was gradually adopted and later became the foundation for the principles of Universal Darwinism.

Darwin's contribution to modern paradigms was to unify and apply the principles of variation, selection, and heredity/retention to explain biological diversity and change. Interestingly, the original theory of evolution was not based on genetics, as Darwin struggled to explain specifically how heredity occurs or why divergence of observable traits may occur in a contextually bound population. Despite this initial limitation, the theory continued to be developed to the form that is widely acknowledged today. Importantly, the work of James Watson and Francis Crick (1953) on the structure of DNA, Stephen Jay Gould and Niles Eldredge's (1972) theory of punctuated equilibrium, Donald Johanson's (1974) discovery of the oldest humanoid skeleton, and Richard Dawkins' (1989) conceptualization of behavior and cognitive habits as the "extended phenotype," and the mapping of the human genome (2000) have extended Darwin's original ideas and provided more detail on the specific mechanisms for the processes of variation, selection, and heredity/retention missing from the original.

---

[3]Dawkins, R. (1976) "The Selfish Gene."

Core mechanisms for biological evolution are now recognized to include natural selection, biased mutation (i.e., difference in probabilities of mutations occurring), genetic drift (i.e., change in the frequency of a gene variant in a population due to random sampling), genetic hitchhiking (i.e., low rate of gene recombination leading to "linked" genes being inherited together), and gene flow (i.e., the exchange of genes between populations and/or species). The outcomes of the process include adaptation (i.e., developmental patterns that enable or enhance survival/reproduction probability), coevolution (i.e., development of a matched set of adaptations, such as those of predator and prey), cooperation (i.e., development of mutually beneficial relations between species), speciation (i.e., divergence of a species into two or more descendant species), and extinction (i.e., disappearance of an entire species). Although these mechanisms originated in biology, they have since been applied in abstract form to explain change in other contexts, such as media studies, cultural studies, art, fashion, technology, architecture, social studies, and economics.

Biological evolution thus forms the foundation for the meta-theoretical framework of Universal Darwinism and helps us comprehend *how* and *why* it emerged. The value of Universal Darwinism is that it can be applied beyond biology and to other areas of change. We can use this perspective to explore patterns in what we now call the "Information Age." We, as a species, have developed complex social dynamics that (enabled by the tools we created) focus on creating, sharing, and refining information. In our modern civilization, we can see that everything can be reduced to the level of information, and it is patterns of information that evolve across different media through instantiation and transformation over time. In this perspective, there is not much difference between genetic information and software languages in how they evolve over time (although of course the specific mechanisms and rates are very different).

What works best at that point in time occurs again in the future. What doesn't work simply does not occur again (over a long enough timeline).

The total information in a certain area (e.g., human genome, Unix technology, etc.) at a point in time is called a "design knowledge base." This design knowledge base changes and becomes more detailed and/or refined over time (a process we have termed as evolution). Some design knowledge bases evolve without conscious human intervention (e.g., naturally formed human facial features), while some evolve through conscious human intent (e.g., Unix technology).

As we have reached an era where information has become the core focus, four distinct types of design knowledge bases can be separated analytically to help explain the complexity of our modern existence: biology, technology, culture, and cognition. As the human species is arguably one that is most reliant on technology for survival (and digital products are the latest in the progression of technological evolution), the next section looks at patterns in technological change that may help us design better digital products for our evolving users.

Technological Evolution

Technological evolution is the rate and patterns of technological change that constitute the developmental passage of the human species. "Technology" in this view is anything that humans have created (intentionally or not) in order to achieve some value. This includes fire, tools, machines, buildings, computers, language, and even memory and voluntary memory recall. Keep in mind that in this view "technology" is not strictly material or external to our biological forms (e.g., a pencil); rather, it is anything that is not directly responsible for the core biological functions of the human body (although, technically, they too can be considered a form of "biological technology" as biological processes are heavily focused on what we have termed in the age of digital as "information communication" and "appropriate timely response").

But before we dive into the complexities of the micro perspective, let's look at technology external to the human body first (as it makes more sense due to our intentional experiences of *using it*). In the last century, we have gone through the evolution of *tools* to *machines* and, more recently, *automation*. The emergence of each signifies major paradigm shifts in global cultural, cognitive, and biological change of the human species (and their environments). In this view, a *tool* is defined as any material entity that extends the potential of human achievement by overcoming the limits of biology (e.g., a knife or axe). *Machines* are defined as complex tools that substitute an element of human effort but require human input for operation (e.g., a boat or car). Finally, *automation* is defined as a machine that is capable of executing commands without direct human input (e.g., robotic equipment).

We can go so far as to view each transition as enabled by previous stages such that as the design knowledge base for technology improves, so does our capacity to create more complex and elegant technology. Curiously, the emergence of each new type does not lead to the extinction of its predecessors; rather, it marks a transition of the human species into the next stage of technological, cultural, and cognitive evolution.

Think simply: we may now have automobiles but we still use knives. Once we developed the automobile, to think we will be driving flying cars next is an example of linear thinking/reasoning. To hope that we get to proper teleportation in our lifetime is an example of exponential convergence thinking/reasoning. The core concept of teleportation addresses the core need (i.e., get from A to B as a cohesive unit/entity) with the current pain points of travel eliminated (i.e., instant transfer over any distance). Although teleportation as a material reality is still far-fetched, we can see how it would hugely benefit a lot of current systems and go so far as to predict that some groups may fight it (had it been invented) to retain the state of current systems that they are reliant on for survival. The challenge of designers is to go far enough in evolving the core without taking it beyond the reach of current state reality (though futuristic ideas in and of themselves are hugely beneficial to the process of innovation because they allow for exploring all possible ways the situation or product *might* evolve and the effect it *may* have if released "into the wild").

Similar to biological evolution, what evolutionary passage a technology takes is largely dependent on whether or not the technology is the most suitable for a certain outcome (or provides specific value). If it is, it remains and other technology gets "added" alongside it. If a new technology emerges that is an improvement upon the previous, then it takes the place of the earlier prototype such that the earlier prototype is no longer commonly used (e.g., think of vinyl records, CDs, MP3 players, iPod, Pandora app, etc.).

Even more curiously, we can observe that technology advances are cumulative. This means that once something is created, new innovation comes at a quicker pace than it took between previous innovations. If we look at recording of written language, there is a larger gap in time between invention of early written media and the printing press than between the printing press and digital media. Within innovation patterns of digital media itself, we can see exponential change and improvement in the last 60 years alone. Along with the rapid evolution (and improvement) of technology, we can also observe very significant changes in global social dynamics, including changes in culture, business practices, social habits, identity, and concept of "self" for a human being. We, as a species, have begun to question our existence in a way previously unexplored: through the interplay between what we now call our "digital self" vs. our "material self." Our sense of self, and its operational identity as "John" or "Mary" has become far more complex than in previous centuries where it was uniquely linked with our biological forms and our social roles in our local communities. Indeed, the concept of community has also changed: we are now players on a global stage, constantly straddling our global (distributed) and our local (biocentric) roles and identities.

How did we reach this level of complexity? Due to the vast volumes of technology we have created to the present moment and the complex history of its emergence, an overview of all documented technological change sequences is beyond the scope of this chapter[4]. However, we can at least note the most prominent technological developments relevant to digital design: those that specifically involve information creation, storage, distribution, and refinement. These include memory (pre-history), language (prehistory), painting/writing media (prehistory), printing press (15th century AD),

---

[4]See *History of Technology* (1976) book series for detailed summaries.

telegraph (18th century AD), telephone (19th century AD), radio (early 20th century AD), television (early 20th century AD), computational devices (mid-20th century AD), satellites (mid-20th century AD), early Artificial Intelligence (mid-late 20th century AD), and, most recently, the Internet and World Wide Web (late 20th century AD). Plotted on a timeline, the pattern of emergence suggests that the rate of technological paradigm shifts is increasing and may be linked to the context of the creators of technology (i.e., accumulated design knowledge), though not necessarily to the rate of their biological evolution. The evolution of technology is much faster than the slow evolutionary processes that constitute our biological forms. We have gone from a paper-based society that used horses as its fastest mode of data transport, to a global network of digital connectivity with 24/7 access to a vast repository of information, in *less than* 200 years.

Futurists, such as Raymond Kurzweil[5], in analyzing these patterns of technological evolution predict that sooner than that we will have another major shift. Referred to as the *Singularity*, the next paradigmatic shift is argued to involve the transcendence of the limits of human biology through merging with emerging technology. It is important to note that this perspective does not assume a dystopian society in which the essence of "being human" is irrevocably altered or lost; rather, the current assumed ontological separation of machines and humans is argued to become increasingly obsolete to the point where it will be impossible to establish meaningful analytical boundaries between the two. Indeed, when we look at a user on Facebook, it is almost impossible to distinguish the boundary between the two except in the most obvious (and least meaningful) way: where the biology ends and where the phone begins. Digital technology, as with all previous technologies, extends the reach of the human being beyond what they can accomplish with biology alone. In this sense, we have already "merged" with technology: albeit not necessarily by literally placing it inside our bodies, but by making it inseparable from us in our daily activities. Technology has been with us, as part of us, the whole time. It is now just becoming more complex and more obvious in its irreplaceable value and significant impact on, not only human life, but life in general.

Kurzweil echoes the perspective of the 20th-century media theorist Marshall McLuhan[6], who argued that media is an extension of man, far more than we realize. His views of the medium of television, and its effect on culture and cognition, suggest that humans have already merged with technology. Not by implanting microchips as early dystopian fantasies suggested, but by adopting and evolving *with* and *through* technology, and using our extended reach to *become more* than what we were without it. Indeed, the push medium of television has arguably provided the foundation for our creation of a uniform

---

[5]Kurzweil, R. (2005), "The Singularity Is Near."
[6]McLuhan, M. (1964), "Understanding Media: The Extensions of Man."

values-based digitally enabled "melting pot of culture" across the world. Yet, recently, we have outgrown the push medium due to recent distributed, on-demand, digital innovations being more valuable for our changing needs than the early medium of broadcast television. Streaming services are rapidly overtaking the home entertainment and news markets because someone else telling us when to watch something, or interrupting the show with advertising, became unacceptable and new media emerged to address our need to consume content *how* and *when* we want it. We evolved, and so did our needs and expectations, and so did the technology that we extended ourselves through. Thus, in a very tangible way, our choices constitute the emergence of "new" and the slow death of certain "old" media (also termed in this book as "legacy systems"), based on what we want and need to do/experience.

Realizing this, we can begin to reimagine our process of digital design and its intended purpose. If we are able to focus on core mechanisms/drivers of human needs, we can design better digital products to meet those needs and help us all to transform into ever new states of being. To do so, we need to shift from being reactive to being proactive in our design. Furthermore, we can start to consciously shape the direction of human evolution through digital product creation and, by extension, of all life on our planet.

To do so, we need to look deeper at these interrelated patterns and start to map out how different media has impacted social dynamics in the past, identify root drivers/mechanisms/needs, and use the technical design knowledge base already acclimated to create things that make life better (whatever the evolving concept of "better" may mean for our rapidly evolving selves). Before we go any further, as designers, we need to navigate the ethics of what we design and the design process itself such that it has no deliberate or accidental negative effects (i.e., those that detract from, rather than improve, health and capacity to independently thrive) occur at any point.

First, we need to realize that *survival* applies to technological evolution in a similar sense as observed in the broader animal kingdom. Survival of the most *fitting* in context. Analyzing the historical emergence of technological diversity, we can conclude that both creation and adoption are very much shaped by the social relations prevalent in a context and value of the technology in the context of these relations. In a very real sense, technological novelty, selection, and replication arise from the interplay of psychological, intellectual, socioeconomic, and cultural needs (among others). Throughout history we can see that technological evolution has been inseparably intertwined with both cultural and cognitive developmental shifts of humans, which have not always arisen from basic survival needs or can be deemed as a natural extension of technological progress.

Think, for instance, of the relatively recent realization of human beings that they may not necessarily be "of nature" in the same way as their garden is. In a very literal sense, they both are (as humans too have been "selectively bred" through culture) and are not (as humans have historically "lived in" a simulated reality that formed as a by-product of the development of the symbolizing human mind). A still relatively common example of an ongoing simulated reality can be seen through religious or cultural practices, business, economics, politics, etc. The question of *who* we are is currently more strongly felt through our social reality, a reality which has little to do with (and has often been at odds with) nature itself. By living in and focusing predominantly on the social reality we embodied and created, we have in a very real way excluded ourselves from other life forms on Earth, and evolved new and more influential mechanisms for evolution that are social, rather than purely biological. Technology is at the core of this artificial (i.e., predominantly cognitive) reality that has long ago become the primary focus of human attention. Think how often one ponders the disconnect/misalignment between one's own logic and feelings, and why the disconnect/misalignment exists in the first place. After all, surely our aim is to be fully integrated and effortlessly functioning as a being. Why then is it *so challenging* to be alive? Perhaps because of the disconnect between life itself and the conceptualization of the human who (for some reason or other) believes that they can control life, govern it, skew it, force it, constrain it, or exploit it. Aside from the obvious ethical issues that are raised during this line of thought, the core question remains: why do/have humans prioritized *anything* over life? How did this come to be? What might we do about it?

Through increasing isolated focus on our social reality, humans have arguably developed a different relationship to "nature" than other living beings. For humans, what is "nature" and what is "artificial" is not clear cut, as we are arguably "natural" but live almost entirely within a reality that we created and actively orchestrate (e.g., the stock market is not directly of "nature" but it shapes behavior and opinions of humans far more than the weather). If left to our own devices, most of us would not survive long in "nature" without the comfortable buffer of technology we have created to distance ourselves from it. Or the technology we have created to, later, bring ourselves closer to it, as we started to see the toll that the artificial reality we began to predominantly exist in had on us: it turned out that our "modern" social reality was in opposition to our own (biological) nature. We realized that we have access to more data and information than before, but we have lost knowledge and wisdom that we had prior that was of "nature." We realized that what we create is always artificial if for no other reason than it being "dead," unable to change on its own without our interference (or us determining its change parameters). We became unsatisfied with that notion, yet unclear on what to do about it.

Applications, virtual realities, chat bots, artificial intelligence, and more have called upon us to further question the distinction between natural and artificial, of human and "other," life and death. They have forced us to confront the idea that, perhaps, we too are no more than complex algorithms running on biology, and brought us full circle to the idea that, perhaps, we have not been of "nature" for longer than we thought. Perhaps we have always lived, to some extent, in a world that we created through various technology, especially since the early days of language where the distinction (and confusion) between nature and our representation of it originated. As at the 21st century, it is impossible to argue that technology is not a driver of global change, as we now live predominantly in a world that does not exist outside our screens and minds. We are further removed from "nature" than ever before, and it is more important than ever to recognize the critical role technology (specifically digital media) plays in our lives, examine how it came to be, and decide where we go next (with and through it).

This may sound and feel pessimistic, yet it is quite the contrary: we are in a more empowered position than ever before to do whatever we want to do however and whenever we want to do it. All we need to do is be increasingly intelligent about what we create and adopt into our existing social dynamics, paying close attention to its impact on health and welfare of all life constituting our planetary ecosystem. We can learn to do so by looking for patterns, learning from the past, and designing for the future we want (rather than focusing on short-term goals, such as profit or social status).

We can begin by analyzing what constitutes the modern lifestyle: what can we *not* live without and why? From there we can look at how these technologies emerged and what specifically made them indispensable to our lifestyle.

Unfortunately to do so is not as straightforward as biological evolution. Unlike biological evolution, tracking of technological change at a micro-level (each individual entity) is difficult as we need a unit of heredity at a suitable analytical level to do this. For example, when comparing two social media apps, how might we meaningfully derive lines of heredity and definitively trace where a specific feature (e.g., the "Like" button) emerged from or its full impact on our emergent social dynamics? Think of two prevalent social content–focused technologies: Facebook and Reddit. Although they both are heavily reliant on the user base for content (and adoption rates), they have two marked differences in design. The first difference is the Up/Down Vote system on posts in Reddit vs. the "Like" only vote system in Facebook. The addition of an extra button (affording users to dislike content or "vote it down") shifts focus from providing the content poster with (arguably) meaningless positive external validation (i.e., if there is no choice but to approve, then there is no real choice at all) to the potential of positive or negative external validation, thus placing in focus the idea itself (and its delivery through written/visual/ interactive text). The second major distinction is the focus on the users' (fictitious or not) self-curated identity. In Reddit, where the identity is hidden and placed on the furthest periphery of users' interaction (i.e., it remains private and only known to the individual user), the content/ideas themselves take priority. In Facebook, where the self-curated identity is the core focus (and core product that users voluntarily create for the company, with murky ownership rights to their self-generated content), interaction is entirely based on who you know who likes what you say/do, rather than on the stand-alone merit of the idea itself. Both digital products have evolved through their user base and feature set over their lifetime, yet still to this day have very distinct evolutionary passages and a very different evolving user base.

So what exactly evolves in the example above? The users, the digital product, the organization creating it, the broader context of humans relating through it? All the above and more. Hence why it is important to look at technology (and evolution in general) holistically. Although the "artifact" has been suggested as the core unit for the study of technology, there is no clear unit of heredity—thus far, technological change was not independent from human beings (i.e., technology has historically been unable to replicate, select, or vary its own design); hence, the unit of heredity has to be transferrable between biology and technology. Technology-specific accumulated design knowledge cannot be isolated in a specific medium or its progression assumed as temporally and spatially bound in the latest sequential artifact instantiation (unlike accumulated biological design knowledge contained in the genome).

To add to the complexity, the recent emergence of immaterial digital media has shifted focus from material tools and/or machines to those that are spatially and temporally distributed and iteratively change in real time (e.g., software, such as the Google search engine). What is deemed as a "technology," or even a "tool" or a "machine," can no longer be confined to definitions based on the concept of a static "thing." For example, "immaterial" digital media, such as software, although devoid of fixed material properties that biological entities can directly interact with, are not devoid of "utility functions" (i.e., properties that evolve over time). Comprehending digital media as an entity is a little like trying to solve the challenge of consciousness: where is it, what is it, and how might we consistently measure and track it? How might we evolve consciously through fully informed choice?

With digital media, to track change, we need to shift focus from chronological technological instantiations (i.e., material artifacts observed at specific points in time) to look at how that technology came to be and all the actors that played a role (direct and indirect) in its emergence. This may include teams of people, product designs, software code, wireframes, organizational structures, user perceptions, and/or development processes. What we need to do is look at how an abstract pattern (i.e., an idea) transformed through instantiations across different media over time, such that it became a digital artifact that then became used by humans in their day-to-day lives. Doing so, we recognize that an app is more than just pixels on a screen, or code, or lists of features, or what it enables and/or constrains the user to do. It is all those and more: the whole is greater than the sum of its parts. Digital media is a composite entity that is instantiated through many different relations across various media and evolves through each. Effectively, we can trace its evolution through a string of decisions. By focusing on decisions that constitute the product's emergence, we can unravel the process of its creation and pinpoint when, where, how, and why the significant decisions were made that later made the significant impact on how the product was used and how it evolved over time.

Let's look at a common industry example. A client approaches a digital agency and asks them to build an online shop. The technology emerging is an "online shop" and by the time it reaches the brief and project sign-off, the technology (a concept still at this point) is translated across multiple people and other technologies, such as Word docs and emails. It then becomes a project plan, a set of user stories, design concepts, low- and high-fidelity prototypes, until it becomes code, and an instantiated e-commerce website on the World Wide Web. By the time real-life users can interact with it, it has evolved so substantially from the original concept that only the core utility function remains true to its first instantiation. Constituted by other technology, the e-commerce store evolves through every decision made, every value and effort of every human (and technology) it comes in contact with. Some design decisions win over others, and some fade away only to re-emerge later on in the process. The final version is akin to a biological being, its process of emergence hidden by its current instantiation, yet, simultaneously, being a true embodiment of it.

This is the true power of digital design. To create something that comes from nothing, exists as nothing, and yet drives human evolution in a powerful way by emerging of us, for us, because of us. Ideally in a symbiotic, rather than parasitic way. It is a part of us, just made of different matter to our biological form. *How might something so different to us be a part of us?* To begin to answer that question, we need to look next to the evolution of culture and human cognition.

Cultural Evolution

The term "culture" refers to a body of socially transmitted knowledge accumulated and refined over time by a species. This includes representational content and the ways that that content is interpreted and created. When we commonly use the term "culture," we, of course, primarily refer to the human species. Although the phenomenon is not unique to human beings, the ongoing transmission and refinement of cultural knowledge has been fundamental to human evolution and our observed disconnect from the flow of other life in our ecosystem.

Cultural knowledge accumulation, refinement, and sharing practices play as an important role (if not *more* important) in our evolutionary passage as biological adaptations. Typically, these evolved as patterns of static attributes that were used to normalize behavior/perception, which were historically used to homogenize/control populations. A simple thought experiment can be used to test the prevalence and rootedness of cultural attributes. Forget, for a moment, that you are an "American," a "New Yorker," a "man" or a "woman," "rich" or "poor," "educated" or "street-smart." You are left with just being "human," same as everyone else in your species. You are life itself, along with all the other species. What would you now do or no longer do given the lack of cultural attributes that defined you before this thought experiment?

It is not difficult to see how "labels" or categorizations (that one, of course, chooses to internalize for one's own personal reasons) determine quite a lot of behavior. To remove them would be akin to becoming a blank canvas, no different in awareness of yourself than existed before we evolved the capacity for voluntary memory recall (a key "technology" in establishing and refining culture), symbolic systems of thought, and, later, an individualistic identity-based sense of self. One would be acting entirely on "instinct" and have no capacity to repeat an action at will (i.e., without a direct contextual trigger). Without this ability, what we now term as "cultural knowledge" could not have been deliberately refined or accumulated. We would only have patterns of behavior that are inseparable from contextual triggers, and the same pattern could be "reinvented" countless times as it would have no way to be remembered and passed on to future generations.

Continuous accumulation and refinement of a cultural design knowledge base has historically increased our evolutionary progress, as well as deepened our comprehension of the universe and ourselves. It did so by giving us a basis from which to operate and build upon, rather than constantly having to reinvent the same knowledge. For example, hundreds of instances of the same design breakthrough (e.g., the wheel) could have occurred if we had no capacity to remember and pass down this information. The value of culture (as our design knowledge base) is that it improves over time, as more and more cultural information is created, recorded, and built upon. Cultural lines are akin to biological lines as a component can, in theory, be traced back to the first instance of that particular pattern. A fascinating example of this can be seen in the evolution of philosophical paradigms and axioms about reality in different human subcultures since beginning of recorded history and their role in shaping our perception of ourselves, shaping social practices, and enabling advances in technological innovation.

The historicity of cultural evolution suggests that *what we are* as human beings is very much a product of culture. It is culture that shapes our constantly changing perception of ourselves, and our ways of being. Arguably, the evolution of cultural elements (both material and cognitive) continuously enabled improvements in our lives across time. Developments in culture have always been symbiotic with developments in technology and human cognition, with culture evolving through us and the technology we chose to externalize and internalize. The more we were able to pass down, the better our cultural design knowledge foundation, the quicker we were able to evolve culturally, technologically, and cognitively. We can argue that our uniqueness can, perhaps, be best characterized by our ability to remember, transform, and pass down elements of information that we can externalize and another individual can internalize with little distortion through patterns of interpretation.

So what is a unit of heredity we can use to examine culture? Richard Dawkins (1976)[7] first proposed the term "meme" to describe a semiotic unit of cultural inheritance. A meme, defined by *Merriam-Webster Dictionary* as "an idea, behavior, or style that spreads from person to person within a culture,"[8] refers to replicating semiotic information units within the "primordial soup" of the whole infosphere that constitutes a culture and the concept of culture more broadly. Memes follow a similar pattern to genes, competing for survival and adhering to the principles of variation, selection, and heredity/retention. Interestingly, memes have at times been in direct competition with genes and vice versa (e.g., meme for celibacy vs. the biological instinct to procreate). Hence why culture is as significant, if not more, than genetic variation in the modern ecosystem. There are few cultures that have not historically placed some kind of social rules on who can reproduce with whom, when, and why. In a very literal sense, we modern humans have been selectively bred by our cultures and historical roles within them.

Unlike biological information transmission, which is linked to a stable and traceable language (i.e., DNA), a unit of culture (e.g., a song) can manifest in a variety of media (e.g., recording, sheet music, live performance, memory, etc.), each of which employs a medium-specific language for information translation, processing, and transfer, and requires different units of analysis at multiple analytical levels. Furthermore, the transfer path of memes across media (e.g., technology and humans) is difficult to trace, as there is no clear mechanism or patterns of actions that can be repeatedly observed as the process of memetic selection, variation, and/or inheritance. Hence, a "meme" cannot be used to create a taxonomy of culture in the same manner as genes can be used to create a taxonomy of biological forms, but it is a useful way to look for meaningful patterns of change in our cultural knowledge base.

Both memes and genes can be viewed as different types of replicator entities, and parallels between them may provide a suitable foundation for deepening our comprehension of cultural evolution. For instance, a "meme" is to the characteristics of an instantiated object (i.e., artifact, idea, thought, song, lyric, etc.) as a "gene" is to a biological organism's phenotypes. By extension of the analogy, individual memes (i.e., patterns of semantic relations) can be linked in a *memeplex* just as genes are linked in a genome (i.e., all the genetic information

---

[7]Dawkins, R. (1976), "The Selfish Gene."
[8]Merriam-Webster, s.v. "meme," accessed February 4, 2020, https://www.merriam-webster.com/dictionary/meme.

of an organism), with the relations between memes in a memeplex constituting the specific phenotypes (i.e., observable behaviors or characteristics). Similarly to genes, the selection of a meme or a memeplex is shaped by its surrounding environment (i.e., multilevel context), which includes other memes being selected. It is important to note that, due to its semiotic nature, memetic evolution is viewed as relatively independent from biology in the sense that the survival of a meme may be only advantageous to itself at the expense of biological evolution (e.g., meme of celibacy hindering/skewing biological evolution rather than directly enabling it, for the entity hosting the meme).

Unlike DNA replication that involves high-fidelity copying, minds or brains alter the information based on the schemata and other interpretation means that a particular human has developed throughout the life of its biocomputer. This process is exemplified by the children's game of "Chinese Whispers" or "Telephone," where initial sequence of words is rarely unaltered through the process of transmission. This is important to consider in digital design as often we are designing products and services that enable translation and recombination of information patterns (e.g., file sharing, messenger apps, etc.), yet rarely stop to consider the cultural implications of these processes: specifically, the common phenomena of misinterpretation and mistranslation. For instance, variation in a cultural context more often comes from replication issues (i.e., copying the pattern wrong) than from deliberate and conscious innovation.

The capacity to represent experienced events in the mind (in our case, in the minds of the predecessors of the modern human being) involved storing the representations in memory and voluntarily recalling stored representations. This process provided the initial conditions in which complex communicative patterns emerged, which formed the basis for cultural knowledge accumulation. The process, in a very literal way, allowed for acts (and experiences they represented) to be remembered, rehearsed, and improved upon. In this sense, memory and voluntary memory recall can be viewed as one of the earliest "technologies" that human beings evolved. To this day, memory is one of the most important aspects for cultural evolution, with many digital products aimed specially at extending and improving cultural memory and reach (e.g., Google, Wikipedia, etc.).

So we developed the capacity to remember events that happened and communicate those events (no matter how accurately) to someone else. We then learned how to share our knowledge of those events to even more individuals through basic externalized representations (e.g., written text). What happened then? As more complex representations emerged through abstraction of events, objects, experiences, and relationships between them, we developed more complex meaning systems. These complex meaning systems and ways to represent them over time culminated in the emergence of distinct cultural paradigms. These cultural paradigms prior to digital technology were localized geographically and have significantly shaped the evolution of technology and, to an extent, biology in those localized areas.

With the creation and proliferation of digital technology, those local cultural groups merged in a global melting pot of cultural knowledge, such that a more or less uniform global culture has emerged with certain local variations based on geographical locales and the history of the area. Digital technology even now is directly enabling the further refinement of our shared cultural design knowledge base and leading to an increased shared comprehension of who we are and where we want to go. It enables us to share more information and compare ways to represent and interpret information. It also allows for us to track in close to real time how cultural knowledge evolves (e.g., forums, chat rooms, etc.).

How did we get to this point? Cultural evolution can be classed in four distinct stages: *episodic culture* (i.e., experiences as representationally stored events in memory), *mimetic culture* (i.e., experience representation classes as actions that can be voluntarily recalled and rehearsed, and passed down through physical action-based communication), *mythic culture* (i.e., development of syntax; cultural meaning and value structures passed down through spoken language), and *theoretical culture* (i.e., development of semiotic structures; cultural transmission based on written symbols and paradigmatic thought). We can observe a distant echo of this process in how modern humans mature through cultural exposure from baby to toddler to child to young adult. It is important to consider this process when designing digital products for different age groups and cultural subsets.

Due to these mechanisms of cultural evolution, we can argue that this process, over time, may have contributed to us becoming further removed from nature. We became caught in waves of misconceptions and ever more elaborate conceptual webs of representations that became our core focus instead of the experiences those representations originally pertained to. The problem is that the representation of something is not the thing itself. It is inherently challenging to create an accurate enough representation so that there is absolutely no ambiguity in what is being represented. It is even more challenging to bring this idea to a wide audience without risk of misinterpretation. For example, if I ask you to imagine a "cat," the only thing I can be sure of is that every reader imagines a different cat. It is likely though that most imagine some kind of furry four-legged creature that meows and purrs. For those of you who thought of something other than what Google Images deems a "cat," take some time to think about *how* and *why* your unique representation came to be attributed to the word "cat" within your cognitive system.

As we can see, a represented experience has many multiples of representation and relies on similar cognitive models (and constitutive experiences those are founded on) for accurate transmission and, hence, transference/propagation across humans. The danger is in creating more and more representational content in cultural knowledge bases such that future generations are led further astray, spending copious amounts of time untangling (or worse, trying to "prove") representations, rather than experiencing life and moving closer to living whatever the "ideal" life may be for them. The danger has been lessened significantly by the content circulating digital media (especially the World Wide Web) as multiple types of content (text, video, images, etc.) could be linked together helping users triangulate representations and get closer to the core of the "thing" being represented much faster and more efficiently than in previous generations.

Due to the nature of culture, when examining cultural evolution, focus on material instantiations alone is not enough—we need to examine how knowledge (and culture) comes to be and that means looking at how meaning is made and shared. Only then can we meaningfully distinguish between the "fake information field" components of culture and those that are aligned with and truly enhance life. Arguably, that is one of the core benefits of digital and core responsibility of digital designers: enhancing life through what we create, and allowing for users to navigate their reality easier and more efficiently/effectively.

Where do we begin? We can start by looking at how cultural knowledge is accumulated, refined, and inherited. We can examine different memes and memeplexes in the culture we originate from (and hence comprehend the most). Memetic evolution can occur through three distinct processes, which can be broadly termed as:

1) copy-the-product

2) copy-the-instructions

3) enhance-design

The process of "copy-the-product" involves mimicking a pattern of behavior or reverse engineering a materially instantiated object (e.g., watching a chef prepare a dish; i.e., mimicry). The process of "copy-the-instructions" involves creating a product from an established algorithm (e.g., a soup recipe; i.e., imitation). The "enhance-design" implies a recombination of materially uninstantiated memes into a materially instantiated form (e.g., combining multiple recipes to create an original; i.e., innovation). In these ways, the concept of a "meme" is closer to that of a "software algorithm" in computer science than it is to a "gene" in biology.

A meme can be viewed as a relational pattern of representational content that is capable of variation, selection, and inheritance, where variation is often (problematically) introduced by copying errors and/or differences in interpretation that are unique to the medium under examination (e.g., divergent interpretations of a symbol structure). A copying error of a meme's representational content (i.e., memetic mutation) would, thereby, introduce phenotype changes in instantiations that can then be mimicked and lead to further mutations (e.g., errors in "copy-the-product" processes) that, if the algorithm is unknown, may become representations without a direct logical link to experienced phenomena the original meme pertained to. As such, this perspective on mechanisms underlying cultural change is similar to Jean Baudrillard's (1981)[9] concept of "simulacra," which is used to argue that copies of copies (i.e., representations of representations) mutate further from the original over time, primarily due to the inability of representational content to capture the entirety of instantiated complexity. Hence, unless the underlying algorithm is known, the "memetic drive" may lead to the spread of representational systems that have no experiential grounding for their content and may inadvertently contribute to the degradation of quality of life and skew the cultural design knowledge base through the spread of misinformation.

How might we analyze how a specific meme in a culture came to be? First, we need to recognize that cultural replicator entities come in two types: semiotic and structural. *Semiotic* refers to meaning, and *structural* refers to how meaning is created through relations between concepts. The adaption of complex mimetic skills is arguably the foundation for all cultures. The process of mimesis involves a direct representation of experienced events and the communication of these experiences through the use of the body, such as tone, expression, movement, signs, and gestures. A cultural example of mimesis is a game of charades or interpretive and/or ritual dance. Complex mimetic skills allowed for the human species to transition from nonsymbolic forms of intelligence still observable in the animal kingdom to a symbolizing mind. These mimetic skills are still a core part of modern cognitive processes, and they too are replicating entities passed down through culture.

Combined with memetic skills, mimesis forms the basis for cultural evolution, and thus we can define two types of cultural replicator entities. The core difference between the two: memes shape the *what* (representational content; meaning), while mimetic skills shape the *how* (relational structures; cognitive schemata). In the digital ecosystem, mimesis relates to structure and functionality rather than readable representational content (i.e., text). Both types move through social groups in waves, inherited through pedagogy and altered through life experiences, where transmission/interpretation differences increase cultural variety and influence replicator entity selection over time. A chorus of a song (a meme) can be traced through multiple media (e.g., sheet music, CDs, live performances, etc.). We know it is the same meme because the notes are arranged in a certain pattern and that pattern maintains its integrity across media. If someone hears the chorus performed by an orchestra and then by a computer, they should be able to recognize the same pattern without difficulty. In this sense, it is a pattern recognition activity at its core.

---

[9]Baudrillard, J. (1981), "Simulacra and Simulation."

To help comprehend how humans operate cognitively, let's delve deeper into how a human being currently gains access to the shared cultural design knowledge base. In childhood, as a human being matures, complex social skills are passed down and imprinted through various experiences. These experiences and the way the child makes sense of them (or is taught to) form the foundation for subsequent cultural inheritance. A child first learns culture based on the information available to them and their capacity to interpret and make sense of that information. They then are able to order it and add to it through changing existing elements as they see fit. Arguably, one of the most valuable aspects of cultural inheritance is the capacity of a child to interpret the information in their own way, thereby evolving the culture at a micro-level. They may then share their interpretation with others and create a paradigm shift if the interpretation is different enough and significant enough in its value.

Hence, cultural evolution is closely linked not only to technological evolution but also to cognitive evolution.

Cognitive Evolution

Cognition has been the fascination of the human species for as long as we have attempted to comprehend who we are and our life purpose. One of the most valuable contributions of the analysis of cognition is the realization that how we think shapes our experience of reality, our choices, and the cultural/technological inheritance we pass down to future generations. The study of human cognitive development involves a large body of knowledge spanning various disciplines, including linguistics, neuroscience, psychology, education, anesthesia, anthropology, biology, and computer science. The term "cognition" refers to the internal processes and abilities that relate to the formation of knowledge, such as attention, working and long-term memory, analysis, evaluation, reasoning/"computation," problem solving, decision making, comprehension, and creation/use of language.

It is important to note that evolutionary accounts of the cognitive development of the human species (i.e., those founded on the principles of Universal Darwinism) retreat from the dualist isolated-mind doctrine, which rests on the central assumption that the mind is "inside" the brain and is to a large extent predetermined. Instead, these perspectives realize that cognition evolves with and through culture. The "outside" (context) of an entity constitutes the cognitive architecture development ("inside") through the outside-inside principle popularized by developmental psychologists. As such, cognitive evolution is viewed as part of a symbiotic relationship of culture and biology and, by extension, technology.

Cognition refers to all processes and cognitive architectures that constitute "thinking," with the brain viewed as the biological processor used to "compute" structured cognitive algorithms that produce the observed effect of the "mind" or "consciousness." When we examine the evolution of cognition, we look to patterns of biological, cultural, and technological change.

Although genetic code provides an important foundation for biological development, context and subsequent life experience play an arguably greater role in cognitive evolution. Genes constitute the basic biological structures that facilitate cognition, but cannot be viewed to inherently produce it—continuous exposure to culture is necessary for mimetic and memetic skills to develop. These skills shape the evolution of cognition of a being. As we evolve, mimetic replicator entities (e.g., pedagogical techniques, rituals, roles, taboos, social relations) form the foundation for subsequent assimilation/accommodation of memetic replicator entities (e.g., language, syntax, semiotic networks, paradigms, ideas, theories, etc.) into a being's cognitive architectures.

As humans evolved, the increasing complexity of social organization and cultural mimetic skills increased development of cognition. The assumption here is that, historically, the ability to develop more complex representational systems provided a social adaptive advantage (i.e., the more intelligent a human, the more they were able to adapt to context and pass on their knowledge). The implication of this assumption is that humans have advanced as far from the rest of the species on Earth as they have due to their capacity to represent experience and operate almost entirely within the immaterial realm of these shared representational systems (e.g., culture). Over time, the increasing complexity of social structures and communicative patterns led to the development of more abstract representational systems, including complex abstract signs, symbols, and syntax.

Studies of modern cognition isolate three major systems core to mental representation in the human mind/brain: permanent semantic memory (i.e., the mental lexicon), episodic memory (i.e., the current text), and working memory and attention (i.e., the current speech situation, external context). Each system is essential to processing contextual information, as well as creating "knowledge" from representational events/types and semantic relations recorded in memory (e.g., mind models). Over time, increases in information storage and processing skills facilitated creation of more elaborate ways to remember and interpret phenomena and enhance predictive capacity for outcomes.

The core mechanism for cognition is the formation of schemata. Schemata organize knowledge and experiential events into small networks of information that become activated by contextual or internal triggers and shape the response of the entity (including processing of contextual input, analysis, and decision making). Schemata are vital high-level conceptual structures or frameworks that organize prior experiences, aid in interpretation of context, reduce cognitive load, allow to predict or infer unknown information, and frame the semantic content of situations. They play a critical role in the development of a sense of self and/or identity of an individual, and the way that sense of self and identity evolve over time. Schemata encompass both mimetic and memetic replicator entities and evolve over time and experiences of the individual.

Schema formation involves two core processes: assimilation and accommodation. The process of *assimilation* involves the reuse of schemata to fit new information (e.g., if a familiar object is observed, it will be integrated into existing schemata of that object). If information is encountered that does not fit any existing schema, then a new schema is formed (i.e., the process of *accommodation*). *Equilibrium* refers to the necessarily temporary stability of schemata, as no new information is encountered that necessitates alteration or formation of new schema. Each process constitutes the development of increasingly complex schema, which allows for more complex knowledge to be internalized, comprehended, refined, and externalized. Arguably, each generation is able to form more complex schemata easier in their biological timeline through the process of pedagogy (as they are able to use the accumulated cultural design knowledge base as their foundation, rather than having to re-create it each time).

As digital designers, we constantly use schemata in design (even if we may not be consciously aware that we are). Schemata are invaluable in reducing cognitive load when dealing with familiar contexts (i.e., development of automatic reaction patterns to stimuli—habits), however, are problematic in dealing with unfamiliar contexts. In an unfamiliar context, many new procedures are constructed or modified, and then tested for effectiveness and the relevant schema changed to the new pattern. Hence, schemata are emergent (not fixed) and constantly evolve through experience. As schemata are the primary way for paradigms to emerge and be inherited, a change in schema is needed for an existing paradigm to change or a new paradigm to emerge. The challenge with evolving schemata is enabling change and (re) formation at a pace that is comfortable for the individual and, at the same time, useful for their functioning in their context. Evolutionary trends of recorded history indicate that "new" has been the adaptive advantage (across all media), while "old" created significant limitations as the surrounding context changed and it limited adaptation to the changes.

Digital media has arguably been responsible for greater and quicker schema evolution than any other medium in recorded history. Not only have digital media allowed for us to become aware of and compare dominant paradigms constituting local cultures and ways of life, but to form entirely new paradigms and subcultures that originated in the digital realm. Digital media enabled us to do this not only through rapid comparison of meaning systems but also in being able to compare how the meaning was put together. For example, Google search engine changed how we think through the very process of using the product. How we learned to access and structure information through using the search engine altered how we organized information in our minds. Social media has altered how we think of social networks and what attributes we value in them. Forums and chat rooms have altered the way that we express ourselves, as well as the complexity and form of language that we use online and offline. Most importantly, digital has fundamentally changed our relationship with the world and with ourselves by redefining our perception of our role and place within it.

A practical example of how designers use schemata is the global UX and UI standards that have emerged over the last two decades. As each digital product entered the market, we as designers learned from their success or failure and each new product we created borrowed design elements and interaction patterns from the previous. Our aim was to improve what we created (even if only a little). Over time, the UX and UI standards that have emerged have enabled users to be able to quickly figure out what to do in each digital product without needing a lengthy tutorial (as was common in the early days of digital technology). This is a practical example of the accumulation of technical design knowledge base of digital products, one which is also most reflective in the simultaneous coevolution of technology and cognitive schemata.

Evolution of Our "Selves"

Based on the meta-theoretical framework of Universal Darwinism, this chapter discussed compatible evolutionary perspectives on the symbiotic evolutionary trajectories of biology, technology, culture, and cognition. These evolutionary trajectories share a common core: the process of evolution is constituted by patterns of information creation and refinement through variation, selection, and inheritance. In other words, we can observe that what evolves across disparate media is essentially *information*.

Through the perspective outlined in this chapter, we can conclude that a significant constituent of what makes us "human" is the way that we create, store, process, refine, and share information. This capacity has been iteratively inherited, refined, and selected within the human species through the symbiotic processes of biological, technological, cultural, and cognitive evolution. We need to consider this in digital design and ensure that we are designing for users using assumptions that are reflective of their state of development, their needs, and their aspirations. We also need to constantly be aware that our design decisions fundamentally alter our evolutionary trajectory and ensure that our designs enable and reflect the future that we want to inhabit.

Biological evolution provides the most historically stable example, as genetic code has limited methods of propagation and can be traced through a common medium. The unit of heredity is a gene (i.e., segment of information on construction and function(s) of a biological entity). Cultural, technological, and cognitive evolution is more challenging in analysis, as the media across which replicator entities circulate encompass both biological and technological components. However, we can still analytically trace these replicator entities across media through two broad types: memes and mimes. Using these in our analysis, we can observe that what we think about, how we think about it, and what we do with this information has largely been a product of our social reality rather than the "natural" world from which we have historically separated ourselves through technology and culture.

As digital designers, we have a responsibility now more than ever to create products that enable us to shift our evolutionary trajectory in the direction we want to go. Digital technology is an extension and amplification of us, of our abilities, such that we are now able to do, think, and feel in ways we could not through biological affordances alone. Arguably, without technology, human beings would still exist in a primitive culture, passing down knowledge directly from entity to entity via mnemonic devices, stories, myths, and rituals. Technology is core to comprehending the passage of human evolution as it simultaneously reflects the development of the human species and enables it—innovation, over time, provides the necessary conditions for further innovation. Furthermore, it allows for us to reimagine who we are, our role and purpose, and consciously shape our evolutionary trajectory.

Digital designers have a responsibility to design products that are meaningful and useful for our users and that also create positive holistic impact. To do this, we need to really get to know our users. So let's jump in and go to the core of modern human beings: let's focus on the development of our sense of self, focusing specifically on identity (a core driver in our decisions and preferences). Who we consider ourselves to be has become inextricably linked to the technology that we use and our reflection of ourselves that we are able to cultivate through it. The next chapter deepens our exploration of how and why we are what/who we are and do that we do, using the focus on human identity as the core driver of the modern human being. We next look at digital users specifically and begin to come up with ways to design digital products that fit seamlessly into their/our lives and create as much holistic value as possible.

Evolution of Identity

The Digitally Enhanced "I"

"I am not one and simple, but complex and many."

—Virginia Woolf, *The Waves* (1931)

When you think of your identity, what comes to mind? Your gender, birthplace, nationality, profession, hobby, the groups you most relate to online, your friends and family? All of the above and more? You are not alone. We, as humans, have developed a colorful tapestry of concepts that we use to define ourselves, to connect with and separate ourselves from others, and to make decisions about who to be, what to do, and what not to do.

Digital products have accelerated the pace of our exploration of ourselves and the formation of more nuanced and unique identities.

What Is Identity?

Arguably, no other species on our planet is as focused on identity and their sense of self than human beings. Our long history of cultural evolution reflects the changes to our comprehension of our self, ourselves and our purpose in

© Anastasia Utesheva 2020
A. Utesheva, *Designing Products for Evolving Digital Users*,
https://doi.org/10.1007/978-1-4842-6379-2_2

the broader continuum of biological life on Earth. How we define and perceive our self and ourselves reflects the changes in our reality and helps us navigate the complexities better and with more meaning. The value of identity is in allowing us to refine who we are, what we do, and how we do it.

Identity, for the purpose of this book, is defined as the set of information that is used to define an entity. Identity is not limited to human beings, but also applies to groups, organizations, movements, fashion trends, weather patterns, and more. One entity may have more than one identity and the same identity may be embodied by different entities in very different ways. Simply, an identity is the set of attributes and characteristics that define "I".

Identity is the set of information that is used to define an entity.

The formation of identity may, but does not require, what we commonly known as "personality." Personality, for the purpose of this book, can be defined as the *way* that the world is perceived, interpreted, and engaged with by an entity, rather than the set of attributes that are used to describe/define it. In real life (rather than abstract models of real life), typically, identity is shaped by personality and sometimes shapes personality. In digital design, personality is a consideration as part of exploring or constructing an identity but is rarely the core focus (as it does not predefine identity). In a similar way, identity shapes behavior and behavior may shape identity, and behavior can be used to trace the actual vs. perceived/reported identity. Typically, personality and identity work together, but sometimes they can stand at odds or even in opposite.

Personality is the way that the world is perceived, interpreted, and engaged with by an entity, rather than the set of attributes that are used to describe/define it.

Identity can be both assigned and enacted. Assigned identity is something that an entity is labelled with that may not correspond with the actions or core nature of that entity. Assigned identities traditionally involve sets of rigid and predefined attributes that are inflexible to adaptation to changes in the context of that entity. An example of an assigned identity is a predefined cultural programming identity that someone is born into and does not choose for themselves, and that may not embody or align with their internal sense of self. Depending on what information others prioritize and assign to a being as it matures in a culture, identity may be based on religion, social standing, sexual affiliation, political views, and more. In such cases, identity often becomes a label that does not correspond to the feelings, behaviors, or preferences of that individual and places the individual's true wants and needs in contradiction to the choices and decisions that their assigned identity affords/allows.

Enacted identities are those that are embodied by the individual as part of their being in the world. They are deeply embedded in every part of that individual. Such individuals can usually be observed to have a much smaller say-do gap than an individual operating using an assigned identity. Enacted identities are more process/values based and have more flexibility to evolve with the needs and preferences of the entity. Typically, an individual may have a combination of both assigned and enacted identities that form a tapestry of what that individual identifies *with* and *as*.

Identity and Technology

Humans have developed identities that are more complex in the 21st century than arguably ever before. We are no longer born into a life of predetermined assigned identity that we cannot or do not question. Rather, we are taught and encouraged to choose our own identity. Invent it, if we will. The core shift of the 20th century was from conformity to rebellion. A significant outcome of this increasing rebellion to tradition, uniformity, and conformity has been a shift from focusing on surface traits to the core algorithms that drive preferences and behavior. We can trace this trend to the availability of information across the world, and the effect of this increased availability on rapid erosion of local culture and emergence of uniform global social dynamics. This has had a profound impact on both identity and sense of self.

The more we learned about "other," the more we stopped focusing on the superficial differences that ostensibly separated us and started to look to the core that all can relate to. We became closer to feeling as one species that is part of the continuum of life, rather than a fragmented tapestry of "us" and "them." Interestingly, the further we delve into that core, the more we realize that we are a part of life itself and not in any way separate from it. The illusion of separation has been significantly undermined during the last century, such that we now have ever thinning barriers to empathy and acceptance of value created by difference. We are also ever more conscious that who we are is a product of our choices, and we are constantly at choice to become something different, something better.

Technology is at the heart of this transformation. It has become increasingly effortless for someone to go into a foreign context/culture and be able to function without difficulty (or even almost feel at home). We can do so because we now have reach to media, language, cultural practices, news, and unique subjective perspectives of local residents who we can learn from before ever stepping foot in their physical locale. Our dress sense has become more unified, as have our social practices, and the way we participate in the global. We can travel further, quicker, and easier. We can spend our money on any product in any currency and have it delivered to our door. We can access countless resources for cheap through mass production and e-commerce. We have become more defined by the "stuff" that we acquire than by where we are born and how we look. We can now look like anyone we want and even drastically alter our biological form to more closely reflect the perception of ourselves we want to cultivate and represent.

We, as humans, have dedicated a large proportion of our energy in the last century to the concept of identity, most recently resulting in the relentless focus on cultivating our own "personal brand." We arguably spend more time now on ensuring the projected image of ourselves gains social currency than ever before. And this process is easier to do than ever before. Part of that process involves tailoring the identity we see reflected back at us through our curated version of our life in social media. Another part of the process is shaping our whole life through the filters of who we think we are and/or who we know we do not want to be.

Coevolution of Identity

In the previous chapter, we looked at four different sets of information that evolve through and with us and shape human evolution and, by extension, all life on Earth—the coevolution of biology, technology, culture, and cognition. Let's explore where and how identity fits into that model. Identity, using this perspective, is positioned at the intersection of the four types of information constituting the modern human. Identity can be conceptualized as the information pertaining to who we are and what we do. In this sense, it is a type of schema that shapes how we perceive and interact with the world around us, how we perceive ourselves, and our role and behaviors that constitute the fabric of social dynamics. Identity helps us define who we are, often by defining what we are not. Some identities are more fluid than others. This is due to the nature of how they are constructed within and by each individual. To comprehend the distinction, we need to examine how we have come to be who and where we are today.

When we look back at recorded history, we see a continuous focus on exploring what it means to be human. Some frameworks have been more prescriptive than others; however, all share the same pattern: be this set of attributes and you are a "this" (whatever "this" means in the local context). Regardless of whether an identity is chosen by the individual to whom it applies or is assigned to that individual by an external entity, the same process of identity formation applies.

Once identity is internalized, unless it is consciously and iteratively questioned, typically, it becomes "hidden"—a subroutine that shapes other internal/external processes. The problem with identity, when it is hidden from the individual who has internalized it, is that often identity becomes a hindrance to the individual evolving and adapting/shaping their context. The assigned identity that the individual internalised, unquestioned, eventually becomes a hinderence to life. Those who are forced to confront their internalized identity early on (e.g., migrants, gay, transgender or transsexual individuals) usually are much better at having a flexible and fluid cognitive system that is based on personal (and highly evaluated) set of parameters they identify with as "them," rather than those individuals who are explicitly/implicitly prevented from even considering such an internal examination.

How does identity form? In childhood, the sense of "I" begins to take form through exploration and interaction with the world. We learn that the combination of sounds others say when they point at us is our name. We learn the separation between "I" and "other" and then develop a sense of self that is distinct to "other." Then we develop an 'identity' (i.e. a list of attributes that we associate with "me"), and we then start to assign meaning to our identity. The identity that started with the feeling of "me and not that" and a name. We then create stories around that name that place us within the social ecosystem in which we are born. We learn "our place" and role in social dynamics. We start to collect and build upon our sets of likes and dislikes. We then learn of complex paths that we can take in life and try them on to see if they appeal to us.

We learn habits along the way and these habits shape the subsequent formation of who we are. If we are raised to believe our value is in our looks, then we are more likely to base our identity on image and appearance than on more intangible attributes such as honesty, integrity, and kindness. If we excel at a sport, we may develop an identity heavily intertwined with that of the sport and the identity of others who choose to participate in it. Our beliefs shape the formation of our identity more so than arguably any other variable. If we believe ourselves to be capable of anything or if we believe that we are capable of nothing, we set ourselves on two very different paths, based on the sets of values and behaviors that constitute the embodiment of those values. Over time, each path enables us to become very different humans.

For humans, the teenage years encompass the last phase of our development as fully formed beings, where identity is most malleable and most formative. The experiences during this time build upon experiences in earlier childhood such that the identity and preferences of a human become more defined and specific to their lifestyle choices and aspirations. There may be links to the way that different humans perceive reality (e.g., visually oriented gravitate to aesthetics and tangible design, logic oriented gravitate to complex concepts and intangible design). The bias that is crucial during this period is focus. Focus combined with the emotional/logical balance and meaning systems enables certain identities vs. others to thrive.

Historically, social identity and biology have been strongly coupled. If you were born a genetic pattern that deemed you biologically "male" or "female," then you had a specific set of attributes you needed to embody and fulfill as part of your existence in the social reality you were born into. If you did not fit neatly into one or the other set of attributes that matched your biology, then you became placed at the edges of social acceptability and your identity needed to be customized by you to fit your unique situation. As time went on, and more and more types of identities and journeys became widely propagated through recorded media, we began to deviate from the archetypal norms that were historically culturally excepted and began to explore ways of being that more matched our ambitions and the life we wanted to lead. New identities emerged through the renegotiation and recombination of sets of identity attributes and the sets of behaviors that these prescribed. This type of identity can be viewed as static attribute based. A static attribute is something that is defined and does not change over time and in variation to context: "I **am** <adjective>_____."

Using this type of identity, something that cannot be changed at will (e.g., genetics) became the constant in life, and its corresponding set of attributes was rarely, if ever, redefined. What it meant to be a "man" or "woman" was set for life (unless consciously questioned, examined, and redefined). As human beings evolved, the concept of identity become more and more pluralistic. What it means to be a "male" or "female" now cannot be predicted in advance. In evolution cultures, a practice of asking what pronouns someone identifies as became the norm, rather than assuming that someone identifies as a "male" or "female" simply based on their appearance and/or biological state. The fluidity of meaning and shifting boundaries that retreated from traditional distinctions now imply that any attribute could mean something different to each individual based on how they choose to be, moment to moment. This type of identity can be viewed as dynamic attribute based. The attributes remain the same, while their definition and patterns of effect change.

"If I **am** <adjective>_____, then I <verb>_____ because_____"

Or

"If I am <adjective>_____ and <adjective>_____, then I <verb>_____ because _____."

It is becoming increasingly important to consider individuals' meaning in comprehending who they are and interacting with them in a way that they accept and deem suitable. We now live in an era of extreme individualism. This individualism has allowed for a new type of identity to emerge. One that is based on algorithms rather than attributes. The core differences between the two are the way that they are constructed and enacted and their adaptive advantages in a given context. The most recent type of identity to emerge in global social dynamics is the procedurally generated identity based on values:

"I **embody** <value>_____, therefore I <verb>_____."

The fundamental difference between these types of identities is that an attribute-based identity does not evolve or change with context, while a procedurally generated identity changes from context to context while maintaining integrity through a common core. The values-based identity departs from reliance on attributes for behavior, instead procedurally generating behavior based on what would constitute embodiment of an individual (or group) of a certain value in a local context.

The final frontier of our lifetime is arguably the decisive choices we make in being who we are and the way that each choice shapes our collective evolutionary trajectory. We flow like an ocean shifting through a multitude of drops. This is a time where we have a chance to experience a new freedom within ourselves to explore what we *can* be, rather than what we *have to* be or what we could choose to be based on our inheritance of all that has come before. Furthermore, through digital enablers, we can now extend and explore our identity in ways that were previously impossible to conceive or even imagine. This process birthed the realization that identity is not static, rather is constantly changing and evolving. The implication of this realization is that we too are not predefined or static but are constantly growing and evolving through different expressions of ourselves.

Given the significance of the impact of digital media on modern identity and our modern sense of self, let's next examine how digital has extended and redefined who we are and broadened the potential of who we can become.

Digital Identity

Identity in the digital realm can be viewed as a spectrum of information extending and defining the individual's self. The most basic form of extending the self is through the creation of a profile or an account. The user fills out a form that contains fields that act in combination as unique identifiers of that specific individual. This combination typically includes such attributes as name, gender, email, phone number, address, date of birth, and an image (either a photo of the user or an avatar). These are static attributes that form the basis for identifying users.

Think back on when you last signed up for an account on a website or app. The fields that were required are likely to have contained this set of information or a subset of it. Depending on the type of account being created, more or less information is required in order for a user to identify themselves and be able to operate in the ecosystem that they need the account to access. Through this common example, we can explore the way that the human is extended in the digital realm.

Over the past two decades, we have extended our memory through online dictionaries, calendars, Wikipedia, forums, photo archives, search engines, and so on. We have curated our own lives through social media. Both personally and professionally, and around causes we believe in. We have found community through forums and chat rooms. We have accessed far more than we previously could through digital content sharing. We have more access to information than ever before. As a consequence, we have become more individualistic. We are under increasing pressure to be "someone." To stand out from the social fabric we constitute, while remaining compatible with it.

We are now free to (re)design and (re)create our identity, perspectives, and beliefs at will, based on a dataset that is not censored or curated by any one individual and/or entity, a dataset that keeps evolving *with* us and *through* us (and us through it). In a very literal sense, we are all more unique now than ever before and also more similar at our core than ever before (i.e., we can recognize that all of us want a peaceful, happy, loving existence, rather than the relentless pursuit of empty attributes and other red herrings that used to distract us and take us further from what we truly want). And by learning how to navigate this increasing plurality/convergence and adapting to, with, and through it, we have become better at comprehending the "root" rather than the "symptoms," giving us better tools and access to our own core, our self—the core of life.

We have access to every conceivable concept and product in multiplicity of customization. In certain countries, we can expect to receive these products after purchasing them online within a few business days. Patience is now something to aspire to, while impatience is what almost every digital product is tailored to minimize, to avoid, to distract from, and to exploit. Some paywall models use the lack of patience (or intolerance to inconvenience and/or wait) in order to extract more money, such as granting access to features, content, and reducing steps required to access it. Other products use identity as a way to control behavior. Most digital products use their features (intentionally or unintentionally) in order to shape the emergence of a specific type of identity or shape behavior of users with that identity, while some digital products remove identity to allow for freedom of speech and reduce bias created through identity. Imagine, for a moment, you are about to share your opinion on the latest political trend on either Facebook or Reddit. Would you post the same thing in both platforms? Why, why not? And how does this relate to the fact that on Facebook your identity is an extension of your physical one, while on Reddit you are anonymous?

The digital realm has been debated since its emergence in the 20th century with regard to whether or not it has significant effect on the social reality that human beings knew prior to the invention of digital media. Given the common place nature of digital in almost every culture across the globe in 2020, we will not debate whether or not digital media has an effect on us; rather, our focus is on exploring how it has emerged and why it has become so significant in our day-to-day habits. Let's start by looking at a common example: the mobile phone. The mobile phone, since the invention of the digital touch screen, has provided a platform through which we have extended ourselves digitally to become more than we were before. In a device that is roughly the size of a human hand, we have gained access to countless spaces that have come to matter more than the "offline reality" that came before it. We now have social media spaces, forums, games, and online communities that allow for us to access and connect with more and more information, and with other humans, most of which we know only through their username and avatar that they select for themselves.

In some digital spaces, we have become quite comfortable interacting with others without knowing anything about them, while in other digital spaces, we have invested countless hours in carefully curating a specific persona/ identity.

The amount and type of information that a user is comfortable providing online varies greatly across digital products and the purpose of interaction. Each set of information is different, though there may be commonalities for an individual across different digital products. For the purpose of this book, we can separate the formation of digital identity into three broad types: *real self*, *augmented self*, and *false self*.

The "real self" is a true mapping of the attributes that someone uses to identify themselves in their material reality and/or social context, such as legal name, photo of their actual face, passport number, social security number, and so on. The "augmented self" is a set of information of which some is a true mapping to their material reality, and some of which is fictional or altered. An example of this is using a real name and location in a forum and using an avatar instead of a photo of themselves. The "false self" is a set of information that someone uses to identify themselves that has no direct relationship with their offline identity.

The reason why a user may choose one type or another is dependent on a number of factors including how trustworthy they believe the digital product/ provider is (e.g., banking app vs. dating app), whether or not this information is required for the service (e.g., government website vs. hobby forum), and how comfortable they are to have their activity online to be related to their offline identity. Typically, a single user will have multiple variations of their identity. Often those identities are multiplicit if fake, and singular and carefully cultivated if matching and deliberately extending their offline identity. In order to provide users a way to comfortably explore their extended self, digital designers need to consider how and why the user's identity is formed/ extended through a digital product.

Identity in Design

In the practice of digital design, identity has been largely implicit rather than explicit. Similar to the way that traditional offline identities have evolved, digital identities started as a set of static attributes and have become more nuanced and complex as users have begun to invest time and effort in the digital realm. That shift has culminated in the proliferation of the practice of human-centered design, whereas digital creators, we first learn about our users, empathize with them, identify their needs, and attempt to create the most fitting product for their lives, identities, and purpose. However, this process is not an exact science, often forcing digital designers to create archetypes or personas out of real-life user groups in order to comprehend their similarities and differences and design appropriately. The focus is on discovering and translating user patterns into useful features that delight the user and reduce barriers to adoption of the product.

Digital designers start by grouping users according to sets of attributes: age, gender, income, religion, and locale. If the project allows, we as digital designers attempt to map different sets of preferences and behaviors and find patterns among the groups so that we can create features and products according to their actual needs and that fit seamlessly into their lives. However, rarely have we looked past these attributes and delved into the process of how our users came to be who they are. Doing so, we miss out on the journey and the key insights that would help us design arguably much more relevant and useful products for our users. When someone interacts with a digital product, they are in a very literal way extending themselves and their reach, to do and be more than they are capable without digital technology. Our responsibility as digital designers is to create products that enable our users not only to reach their goals but also to benefit from using the product. If we forget that that is our responsibility as designers, then we run the risk of creating products that negatively skew the way the user evolves, especially in perceiving and cultivating their sense of self and (re)forming their identities.

How a digital designer selects the elements of identity, and its representation and embedding into habits created/reinforced by the product, depends on the ecosystem that the product is a part of and its role within it. Products that are utility focused use identity of the user as a background identifier, needed primarily for the knowledge that you are representing this "person" or "entity," and are hidden away from the user in their interaction with the product or service. The core focus of identity in these products is on the identity (i.e., branding) of the organization creating the product. An example of this is a telco or government service site.

Other products that involve communities of distinct users interacting within its boundary have an equal weight on the identity of the users and the identity of the organization and/or product. Examples of this are forums or sports apps.

Other products reduce their own branding (at times to the point of removing it completely) in favor of focus on the identity of the user. The digital product in these cases merges seamlessly with the identity of the users. The users adopt the functionality and aesthetic characteristics as an almost literal extension of themselves. Their identity becomes framed and shaped by the digital product, almost like a work of art. Neither exists independently after the initial merger and the more the online identity is perceived as an extension of the offline identity, the more time and effort the user is willing to invest in cultivating it. Who we see reflected back at us becomes more tailored and evolved over time to be more fully "I", and the reaction of the user to the reflection they see often shapes the next set of identity attributes the user posts and the practices they engage in online and offline. An example of this is popular social media platforms.

Parts of identity that are not considered by the user as "official" or "public" parts of the self also gain the opportunity to be explored unconstrained in some digital products. The value of adopting a fake identity is the reduced risk of having a part of the self explored, a part that may not be palatable or advantageous to the user's offline identity. There is a sense of freedom in the digital realm due to the possibility to be someone other than who you are considered to be offline and the low risk of these explorations on disrupting the trajectory of the offline identity. Such explorations enable the user to develop a personal scale of identity and a more complete and nuanced sense of self due to the freedom to explore parts of themselves that they may otherwise not have the opportunity or courage to embody.

The digital realm enables us to deviate for a time from what others perceive and know us to be (and what we perceive ourselves to be) in the offline reality and become the multiplicit digital "other selves," "avatars," or "alter egos." From our exploration, what we deemed as palatable then may be integrated into the offline identity and become a shared extension of the self. What we deem as unpalatable can be exonerated into the shadows of the digital world and removed from any relation to our existing offline identity. At times, the two may crossed paths, like secret lovers pretending not to know each other or feel a kinship to the same inner core. The value of this aspect of digital is the potential to explore all that we are and take what we need and discard the rest. This exploration allows for us to realize that what we are is not and cannot be contained by the reductionist view of ourselves we may have had previously. Rather, we have gained the opportunity to expand ourselves, explore, and encompass the depth and complexity of all that we could and might want to be.

Because of our constant evolution and the pivotal role that identity plays in our lives, in digital design, we need to be mindful of how we extend the user and the impact(s) that our designs may have on their lives. We need to know how to design for an evolving entity and know what to focus on in the sea of noise that the modern context can be. To develop our design skillsets for this purpose, we will next explore in detail how to design for evolving users (and their evolving identities).

Designing for Evolving Users

Enhance Life Through Digital Design

"Life is about creating yourself."

—Unknown

When we set out on the journey of creating a new digital product (or revamping an existing one), we are often faced with an eerily blank canvas. The potential for creating *something good* (and aren't we always striving to create something good?) can be as scary as the potential for creating something difficult, meaningless, useless, or just plain bad. The uncertainty usually comes from the realization that the difference between the two is ultimately decided by the end users, and this means one thing for those involved in the creation process: we have absolutely zero control in how someone will use and/or perceive the fruits of our labor once it's in the wild.

Knowing that the fate of the product is predominantly in the hands of humans we have not met and are unlikely to ever meet can be a daunting realization. Seasoned designers have learned to put personal preferences, ego, and preemptive claims to fame and fortune aside and have become powerful conduits for the ideas that come through us during the design process. With our purpose as translators, curators, and orchestrators in mind, digital

© Anastasia Utesheva 2020

A. Utesheva, *Designing Products for Evolving Digital Users*,
https://doi.org/10.1007/978-1-4842-6379-2_3

designers have assumed the unique role of a bridge between the intangible and often unspoken, and the tangible and deliberate value-adding.

To deal with complexity, we, as digital designers, have not only learned to swallow our pride and make decisions that benefit our end users (rather than ourselves or those commissioning the product) but also learned to navigate the limitless potential for feature combinations, shifting user wants/needs, and the pressures of commercial realities of bringing a product to market. As digital designers, we constantly ask:

- Who are our users?

- What needs are we addressing?

- What is the purpose of the product?

- What features are most important?

- Why would any user want to adopt our product (given that they are probably already inundated with similar products)?

- Can we do it a better way?

- How might we create the best possible experience?

- How might we stay relevant and consistently add value?

As part of this ongoing quest for relevance and desirability, we have come to realize that we, as digital designers, have the unique opportunity to create something that enhances the lives of users and helps them to navigate the increasing complexity of our ever-evolving shared reality. So how do we do it *better*?

In the previous chapters, we explored the context of digital media and its role in the lives of humans and the ecosystems they constitute. The core takeaway from the exploration is that in any media, what evolves is essentially *information*. When we examined the evolution of information across media, we came to a focal point: the human being extended through technology. This is the point where we can observe in real time the coevolution of the four core types of information (i.e., biology, technology, culture, cognition).

When we design a digital product, we always have these types of information in mind (however unconsciously). Usually this happens through the necessity of making something relevant and desirable for someone else. The value of unpacking the types of information we are dealing with and helping evolve through the digital product is the ability to focus on what is important in a sea of noise that we need to navigate at the start of any project:

- What parts of culture are we building upon and/or challenging?

- How do the users expect the product to behave?

- How will the users know what each feature does?

- How will they find/discover said features?

- How much will we need to explain before the user can successfully use the product?

- How much of the user's own unique journey/preferences are a barrier and how much are an enabler?

- How much value does the product actually add to users' lives?

Our laptops and mobile devices are the literal culmination of the last two decades of experimentation in practice with regard to digital products and services. These last two decades have yielded a significant body on knowledge of best practices, digital no-nos, and lessons learned that we can collectively benefit from in our projects. This chapter is intended to provide a high-level summary of the highlights with the intent to help us, digital designers, to design better, quicker, and with more impact.

Being a Digital Designer

Let's dive into the core of digital design. When we design digital, we go beyond designing a material "object," one that is unchanging after designers create the blueprint and that blueprint becomes a replicated "thing" out there in the wild (i.e., beyond the control of the original creators). As what we are actually designing is an interactive experience in which all elements participate as co-creators of said experience, we need to be aware of how what we design might be interpreted and what each possible interpretation may mean for the users' experience. Think of it as creating a choose-your-own-adventure story, of sorts. With many actors that come and go as the story goes on.

When we design digital, we go beyond the traditional focus of design such as balance, pattern, contrast, color, texture, shape/form, emphasis, repetition, proportion, rhythm, variety, and unity. Yes, we use all those as part of the design process, as we are still designing elements of a tangible and aesthetic product. In digital, the core focus, however, is on the *experience* itself. An experience goes beyond the traditional boundaries of a "thing" that someone can "use." An experience is a sequence of events in time, often composed of multimedia and multisensory information. It also requires active co-creation by the users in order to happen. The role of the user is active (not passive), and we often cannot predict how they behave within the parameters of what we create. We can only design the affordances and constraints that create the conditions for someone to be able to do something. In other words, our role is more akin to a magician than a blacksmith. Leave clues and hope the users can follow the trails we lay out for them.

Like the old saying "you can't make someone have an epiphany, but you can create the conditions for the epiphany to occur," in digital design, we are creating the conditions for someone to do what they want and/or need to do, though we at no point can or should try force anyone to do anything. The possibilities that may and do happen arise through the simultaneous interplay of the affordances and constraints of biology, technology, culture, and cognition. Though, saying that, we often find ourselves helping users stay on the path of competing a task and preventing users from doing it completely the wrong way (e.g., think of Xero accounting software or the game of Solitaire on your desktop, which have intrinsic rules built into them).

The tricky thing about creating a set of affordances and constraints in the digital realm is that there is a lot of potential for interpretation and action that is not intended. This is because *anyone* can use the product (if it's public, of course). "Anyone" means all humans (and even now some other species, such as iPad apps for dogs, chimps, and dolphins), in all the forms that they are in at the time of using the product. Lots of different interpretations and intents lead to a variety of user journeys and pathways.

Let's stop here for a moment and examine the word "using." When we *use* a traditional tool, like a needle for embroidery, we are literally extending our reach beyond what fingers (and other components of the human) can do alone and creating an occurrence that is not directly "of nature" in the same way that fingers themselves are. The shape and other qualities of the needle itself become what they are because of the limitations of human fingers and the intent of the human to do something that biology alone could not accomplish. So the human extends the reach of their capability through the "tool" of the needle. However, when you watch someone do fine needle work, it is difficult to meaningfully separate the needle from the fingers, the fingers from the body, the body from the mind, heart, and soul of the individual. The needle is no more creating the end product of an embroidered pattern than the fingers themselves. The creation begins at the point when the human decides to create something and recruits the necessary extensions to themselves in order to make it a reality. Not the other way around.

In digital design, a similar phenomenon occurs, except that it is even more difficult to distinguish between the "user" and the "tool," as we are extending the intangible elements of the human (e.g., capacity to remember, to think, to analyze, to communicate, etc.). In essence, to "use" a digital product is to "merge" with it and become one as part of the interaction in the most abstract and unpredictable way. It is harder to predict how someone might interpret our app than how someone might interpret the chair we create. In saying that though, the more that someone has exposure to digital products, the more likely they are to reverse engineer our logic and even see beyond what we originally intended in the design process. As such, when we create an experience, we look to the highest possible point of abstraction (i.e., what pattern is the most core and the simplest way to enact it). The more abstract, the more open to interpretation. The more open to interpretation, the more potential. The more potential, the better the chances that someone might find the product useful (or get overwhelmed).

If you think about it deeply enough, the word "using" in the context of digital media is a bit of a red herring because in digital design we go beyond an extension of one capability which is implied by the word "using" and its companion "tool." What we are actually doing is creating a *space* for someone to do what they cannot do external to it. Setting up a playpen with certain parameters, so to speak. Those parameters form the boundary of the pen and, by extension, of the experience. What happens within those boundaries, neither the designer nor the user can predict. It is a process of unfolding in real time, a *happening*, that requires equal input from the user and from the technology itself in order to occur.

One of the main pitfalls of digital design, as those in the industry can attest, is that it is often tempting to funnel the user, to force them down a linear predefined path, so that we can confidently say that we (as designers) can guarantee we deliver on a certain X or Y objective and/or outcome. But to actually do so would be a complete disservice to our users, and that's why we resist the urge (and external pressure) at all costs. As anyone who has owned a needle can attest, it is incredibly useful for the attainment of all sorts of outcomes, not just of sewing or anything that involves a needle *and* a thread.

For example, a needle can be used for freehand embroidery (where you create designs from your own creative intent). The same needle can also be used to create a predefined pattern (one that someone else came up with). A needle can also be used for entirely purpose-attainment reasons, such as using the needle to hem a sleeve. The needle can also be used in a completely different way to form a teeny tiny spear for the fairy in your garden.

The freedom to use the needle past its intended purpose (as either the vision of the original designer or the manufacturer of a modern needle) is at the core of why it is a valuable thing to have in our homes. The potential of what the needle can do is both a gift to and a right of every user of the said needle. The limit of the potential rests with the limits of our imagination.

The same applies for the digital realm. We cannot and should not attempt to constrain a user and force them to do things inefficiently or ineffectively, or just plain uncomfortably, just because we designed it that way. Here is where the responsibility of a digital designer is felt most. We are creating something to make someone's life better, not worse. And to know how to do that, we need to get to know our users, to get to the core of what drives them, and do everything in our power to help make their life better. That is why, half the time, we, as digital designers, seem to be advocating the rights and needs of our users more than designing the product itself. And that's a good thing. Because if we take on the responsibility of design, as part of our role, we need to know our end users best. To *be them* if we can (or come as close as possible). Empathy is arguably the most important skill in our design toolkit (see Chapter 4), and there is no greater empathy than that of a shared experience. More often than not, the process of design becomes a quest of selfish altruism. At some point, we always realize that we, the designers, will likely be using the product ourselves. That is why the gut (a.k.a. "sanity") check is another important feature in our design toolkit, and some of the most elusive insights often come from trying out the product ourselves to make sure it is fit for purpose before we inadvertently inflict our creation on others.

The example of the fine needle work illustrates the core of what we need to get to in digital design in order to be successful: *we need to extend the reach of the human beyond what they can do currently and in such a way where they (the users) have complete control of the experience and its outcomes (in a guided and pleasant way).* What we design when we design the digital product is *potential.* Yes, we also design an extension of someone's limits to enable them to achieve what they need and/or want, but that is simply a *means*, not an end.

That's why what we design when we design a digital product is a much more substantial extension than a limited material (i.e., unchanging once made) "tool" such as a needle. What we are designing is an extension to the mind. The mind can be thought of as the software that runs on the hardware of the physical body. Schemata in this view are software programs that allow the user to perform higher-level processes (e.g., speech, conscious thought, memory recall, etc.). The hardware in this analogy is the physical body of the human (and all that is also in the same tangible material realm as the biological human body). What we are extending in digital design, first and foremost, is the *mind*.

As a symbolizing being, it is arguably natural that, once we figured out how to, we would extend the reach of our minds beyond what nature was able to evolve as the state of the art of the biological form. We can observe the same patterns in recorded thought (i.e., illustrated or written texts of prior generations), as we can in the evolution of the World Wide Web since its inception. The eerie parallels between the two are worthy of contemplation. The wonderful aspect of exponential convergence (which we have mathematically established is a rather accurate reflection of reality) is that if what evolves is information, then we as digital designers are at the bleeding edge of evolution of life itself.

A word of warning: when we get to levels of abstraction in thought (and in design) where we are considering all of life, then it is inherently dangerous to be human centric. The reason it is dangerous is because of the biases that we have inherited a part of our social fabric of reality. Biases, that until they are resolved, may be skewing evolution toward human legacy systems concepts such as profit, exploitation, or enslavement (rather than the pure aim to enable life to express itself as life, and only life, wants).

To be a digital designer, you need to *know yourself*. And that is why designers generally prefer to collaborate in teams. If you are bringing in unconscious bias to the design process, having a team of humans that think differently adds unquantifiable value. Different cognitive systems coming together to achieve an agreed-upon activity allows the differences and similarities in patterns of thought to be noticed early on and analyzed. It is important to note here that there is nothing inherently wrong with bias—if I were to hire a designer, I want them for their unique perspective and skillset (which is effectively a form of bias). The kind of bias that we want to reduce is the kind that is due to *personal preferences*. This kind of bias gets weeded out pretty quickly in a team of individuals with unique cognitive processes, as all bring a unique perspective in the process of collaboration united through the focus on something much bigger than each of them in isolation.

Sometimes as digital designers, we almost feel like we are reaching into the future. In fact, what we are doing is *creating* the future. And the creation process for a specific future is multifaceted. Our aim as digital designers is to create products that enhance our collective future, rather than degrade it. How might we do that? There are a few basic digital design principles that may help.

Digital Design Principles

1. Design is a conversation, not a monologue

Design is always a conversation. It is never a monologue, even when it is a sole endeavor. There is always an exchange of ideas that happens as part of the process, even if those are ideas generated within your own mind. It begins when a designer comes up with an idea or gets a brief. The designer then starts the internal dialogue within themselves about how to go about designing what they set out to. This dialogue is informed by all that they know and have experienced, and the quality of which depends on the quality of the "software" of rational thought of that individual. Then the conversation extends to *other* (thinkers dead and alive, personally and publicly known). This may include a team of other designers who have adopted the same design challenge, the end users, those affected by the end users, those who commissioned the product to be created, and even those we meet through serendipity along the way.

As they say, a good idea can come from anywhere. Those of us who look for inspiration outside the confines of the expected typically come across the best insights.

Design with the intent to bring the end user into a *constantly evolving conversation* with the product and with others who share that space.

2. Start with users' needs and wants

When we create a digital product, we do so with an end user in mind. That user may be a human, another living being, or another technology. In order to design affordances and constraints of the digital product appropriately (i.e., as best for all involved/affected), effectively (i.e., elegantly), and efficiently (i.e., effortless to engage with), we need to know our users. Get to truly *know them* (remembering that we can only truly know *other* when we truly know *ourselves*). Help them to get to know themselves better (even if it's only a little bit). When you ask someone to tell you about their lives, rich tapestries of perception and interpretation become revealed not only to you (someone hearing them for the first time) but also to the one doing the talking. The skillsets to navigate and accurately represent these stories are core to each designer's toolkit. We'll talk more about the toolkit in the next chapter.

There is a qualitative[1] difference between needs and wants. *Needs* are something that arises out of necessity (i.e., cannot happen without), while *wants* are more elusive. Wants are arguably what we truly design for, because if something is wanted, it is a reflection of its relevance, if not a subjective measure of success. And because it is subjective, the only way to know is to ask and listen carefully to the response.

3. Make it simple

Creating something good, something elegant, requires simplicity. Simplicity is when we look through the noise of all information and all its potential and find the way that is most *convergent*. Most convergent means that something is universal enough to be applicable in its form to all (or as many as possible). An analogy for this: we all know Shakespeare wrote about *love* in *Romeo and Juliet*[2]. He did not at any point literally state in the play: "I am now going to write about love" and then define "love" or describe how he (i.e., William Shakespeare himself) experienced it. Nonetheless, we, as the audience, can come to the conclusion that he did in fact write about "love." We can do so

---

[1]Oxford University Press defines "qualitative" as "relating to, measuring, or measured by the quality of something rather than its quantity" (https://www.lexico.com/en/definition/qualitative, accessed July 20, 2020).
[2]Shakespeare, W. (1597), "Romeo and Juliet."

because we can see patterns in the text that correspond to patterns that signify "love" in other texts and, of course, through the patterns of personal lived experience of what each of us knows as "love."

Converge, converge, converge. If it is not simple enough for someone to pick it up and instantly (or with little strain in thought) make it work for them, then it is *bad design*. Good design doesn't need explanation. It just delights when encountered and makes us wonder in awe of *how* and *why* we didn't think of it before!

True simplicity is, arguably, true beauty.

4. Build on what works, innovate as needed

Designers from any discipline will giggle quietly to themselves when they are asked to create something "new." The basic practice of design requires us to always look at what's been done, take the best, figure out why the rest doesn't work, and iterate, iterate, iterate, iterate, until we have something that is significantly better (whatever "better" means in context). That is how progress happens. By evolving what came before. That is also how innovation happens and how we achieve true innovation[3] (not just a repackage of existing patterns). By necessity.

5. See everything through patterns

Everything is information. Information forms through waves of patterns[4]. See all reality through evolving iterative patterns, and you can start to design patterns. Which is effectively what anyone is doing when they create anything (especially digital). Once one learns to consciously see patterns (any and all kinds of patterns) in all in real time, navigating reality becomes much easier. Not to mention to design relevant and meaningful patterns, which is the primary activity in digital design.

Analogous thinking (and the inspiration that comes of such thought processes) is key here. When you can see the same pattern manifest across media, then you can figure out the mechanisms, triggers, and outcomes in a systemic and holistic way, and then design for them or design for/to change.

---

[3]"Innovation" here refers to something new that adds value. It is purposeful (in terms of being value-driven), unlike pure invention that can exist only for the purpose of it existing.

[4]"Patterns" here refers to any arrangement of information. When abstracted an arrangement of information can form another arrangement (one that represents the previous arrangement). If the abstracted pattern can be applied to multiple arrangements on a lower level of abstraction, then that arrangement can be called a "meta-pattern."

6. Be consistent

Consistency is achieved through repetition of patterns. If something is consistent, we can predict what happens next based on the patterns that came before. In a procedurally generated cognitive system (as outlined in Chapter 2), consistency comes through the algorithms that create each point of interaction. When patterns are similar across media and space-time, then there is consistency.

Consistency applies to reusing elements across the entire interface and to the patterns of interaction itself. The more consistency (not "sameness," *consistency!*), generally, the happier the user (it reduces their cognitive load and helps access core information better).

7. Iterate, iterate, iterate, iterate, then iterate some more

If you try to design something *once*, you will find it extremely challenging, if not impossible. Try to design a paper airplane once. You only get the first impression that comes to mind (as your final design), because if you change the way it looks or works in your mind, that is a form of iterating the design. Ideas have this habit of changing quickly, recombining on the fly, and transforming themselves in real time. One of the best activities to enhance creativity is in any ideation session to design something at least five times, ideally in different ways. Preferably, in analogous ways at first, and then on an actual convergent design that you can see the product forming through. As designers, we always iterate multiple times throughout each design activity and countless times throughout the duration of the project.

Ideation involves creating many ideas (diverging; through a focal concept or set of parameters), then converging those ideas into one (or a few) idea, and repeating that process until you have something that is one idea (and it's the best one yet).

The value of iterating often and from the get-go is that the ideas evolve quicker and, thus, product becomes better quality. Ultimately, iterating more exponentially increases your chances of "success" (whatever that means in context).

8. No fear, all the curiosity

As designers, we don't really fear much. Not our successes, not our failures. Designers have been so defiant of the historical concept of "failure" as to redefine it and its value completely. Designers see failure as growth. That means that at no point have you or anyone else "failed," but you and everyone else have gained a lot of knowledge and skills in the process (and possibly found a few ways that don't work). That's why designers are so fond of iteration. If you do not "fail" in the first couple of attempts to create what you set out to, you have not started the creation process. That's where the concept and process of "iteration" gains its true value. As part of the holistic process of design, we always try things, observe what works, analyze why/why not, come up with insights to refine the design, and create ever improved designs (evolutions of the previous, yes, but still improvements).

The beautiful thing about first finding what doesn't work is that you get to over time truly hone in on what does work and why. In ongoing iterations, failure is small, manageable, constant, trackable, easily identified/analyzed, and constantly learned from *before* the next iteration happens. This not only improves quality dramatically but also allows for concepts to be tested quickly and easily and refined before the design is deemed as "complete" (arguably, no design is ever complete, but every good design becomes at some point production ready and released into the wild).

Curiosity is one of the most valuable qualities to nurture because it allows for us to pick at all the relevant threads, without fear or judgement, until we gain a deep and profound holistic comprehension of the phenomenon. It also allows for us to seek out failure because, like in science, failure provides designers with valuable insights as to *how* and *why* something may work or not work. When there is abundance of curiosity, truly innovative ideas/designs thrive. Curiosity works best in conjunction with empathy.

9. Look beyond the surface

A surface comprehension of anything ultimately leads to a bad design. Or, more accurately, the first couple of iterations of the design reveal the gaps in comprehension that need to be resolved before the next round of design can occur. The first couple iterations of designs are always "bad designs." That's their whole point.

Beyond every great idea or insight or epiphany is its own logic and its own historicity that make it unique and tell a story about *how* and *why* it came to be the way it is. These stories allow us to lift the idea out of its history and place it somewhere new.

Look beyond the surface, and design from the core.

10. Transcend the now, create the future

Every design must be an improvement upon the previous. Otherwise, there is no point to the design at all. To design without improvement would literally be us doing the same thing, and that would create a very monotonous (albeit consistent and predictable) experience. Besides, it is illogical to think that there is nothing to improve, or think that maybe everything already exists, and, hence, it has all been designed already.

Design for the future you and your end users *want to* live in. Do not ever settle for anything less and always do your best to create the design such that the affordances and constraints of the digital product *enable* that ideal future to come into being.

And finally, the meta-principle.

11. My life, my way

Design with this last principle always in the front of your mind. Got it? Good.

Digital Design Process

The process of digital design is abstract, messy, and inherently fun to do. Like in any design process, we first get the design brief (i.e., a set of parameters that form the boundaries and often the success criteria for the design), form a team, and set out on a journey of exploring, analyzing, ideating, trying, and learning.

To design in multidisciplinary teams is to design better. The convergence of different perspectives gives us an advantage by (paradoxically) weeding out "groupthink" and maximizing the impact of design. When a marketing specialist, an accountant, a fashion expert, and an art curator come together to design a new app for a fashion exhibit, likely the things that they can all agree upon will be the core to design and the individual perspectives they bring to the table the source of the most valuable insights. Seasoned design teams often bring in polar opposite perspectives and skills to the core design team for this very reason: healthy conflict (i.e., conflict of ideas and perspectives, not personalities and agendas) results in a more robust and *creative* design.

The design process itself is nonlinear, and often the "breakthrough" insights come to us at the most unexpected of times and places. This is why designers are often found in cafes, parks, or wandering around looking for something to play or prototype with. Unlike other skillsets, the designer's toolkit is often composed of more of methods and approaches we are trying out, rather than those we have spent years refining. The process that we adopt in order to design needs to be flexible enough to allow for ambiguity and complexity to run their natural course, but also be structured enough to enable us to adhere to the commercial realities of the commissioned project.

How the design process actually unfolds typically depends on the overall methodology adopted by the organization and/or project team. In the early days of digital product creation, "design" was a word that was rarely (if ever) used to describe the process. Typically, we were "building" or "developing" products. Traditional Waterfall[5] approaches were very linear and involved months, if not years, of requirement elicitation and analysis and a long development process. Users were often consulted in the beginning, if at all. The end product often was more of a constraint than an enabler as organizational processes and habits had to change in order to fit the technology created (rather than the other way around).

The linear Waterfall approach had its obvious downfalls, and more flexible Agile[6] approaches came along in order to help digital designers iteratively create the product in short cycles of improvement starting with the MVP (i.e., minimal viable product) and cumulatively evolving it in response to arising and changing needs of users. Agile approaches were a significant improvement upon Waterfall approaches as the design component of the process became more prominent and users were invited earlier on and became a stronger voice in the process. Designers were able to do research (rather than requirement gathering) and dig deep into the problem/design challenge before any product was conceived. The change in approach also reduced the pressure on the team to come up with the perfect design on the first go, by allowing for iteration, and this greatly improved the quality of products created.

The most recent approach adopted by digital design teams is the Design Thinking approach, which originated in industrial design and has been pioneered by organizations like IDEO in its value and application to all types of design challenges. This approach goes further in emphasizing design than Agile approaches, as the entire team become "designers" (rather than product owners, developers, analysts, researchers, and users) and all have equal voice and responsibility for the entire product. Design Thinking has also yielded the best results for incredibly complex and abstract challenges (e.g., reducing inequality or improving health and well-being in marginalized communities), as the whole approach is based on deep empathetic comprehension of those affected. Unlike Waterfall (and even some applications of Agile), the end users are not expected to dictate the solution. Rather, they are asked to share their experience and, together with the designers, are invited to explore their challenges, thoughts, and feelings and come up with ideas for improvement.

---

[5]Linear approach to software development, with sequential phases of conception, initiation, analysis, design, development, testing, deployment, and maintenance.
[6]Iterative, nonlinear software development where requirements and products continuously evolve through the collaborative efforts of multidisciplinary cross-functional teams.

Design Thinking and Agile development methodologies are arguably the most valuable approaches for digital design, as they allow for and enable evolution of the ideas such that they get better over time and more closely aligned with user needs and changes in context.

Designing for Evolving Users

As digital designers, we have the opportunity to help (re)design the modern self and the way of life that constitutes it. When we design a digital product, what we are actually designing is a new way of life, a new future for all. We need to keep that in mind at all times. That is how we prevent ourselves from dissolving into obscurity of linear incremental change.

In the 21st century, the exponential rate of holistic change is difficult, if not impossible, to ignore. The exponential rate of change in general suggests the end users of our products are also exponentially changing. They are transforming through their ongoing interaction with other digital products and the ideas they circulate. To design a digital product with assumptions of a static nature of users or assumptions of unchanging needs is to set out on a path of creating a product that is relevant only for a very short period of time (if it is relevant at all by the time it gets into the app store). Hence, why any good design in our modern context is a design with the concept of evolution (and thereby exponentially) at its core. The products that enable us to evolve better and in the direction we want to go are the products that we keep near us and refer to often. They remain relevant because of what they provide for us and enable us to see, think, feel, and do. Often these digital products evolve with and through users (e.g., Google search engine, Wikipedia, etc.).

How might we design something that will evolve with and through users? Let's return to the focal concept of Chapter 2: "identity." We defined "identity" as a set of attributes used to define/identify something and/or someone. We then explored types of identity and defined three broad types: static attribute based, procedurally generated, and composite. To design for the modern human, we need to design for *both* static attributes and procedurally generated behavior (based on a core set of values). The reason identity is so prominently highlighted in this book is because it is a core filter that the modern human navigates reality through and, as digital is an extension of self, we need to know how identity fits into the picture so that we can design truly meaningful and useful products.

To design for an identity, first we must become curious about what kinds of identities are out there in the user group, triangulate on the similarities and differences, create personas (not the same as identity), and try to figure out how to make our user's life better. Start with the most basic scaffold of an identity first, then build it up as needed. Don't collect more data than you need to and always, always respect the privacy and decisions of the user. Their data is (arguably) theirs, no matter who owns the server.

Be careful when asking users to self-reflect or self-curate. Most humans, unless they spent time learning the art of self-reflection and reflexivity, will struggle to give you the answer when asked point-blank. Most will need a scaffold of questions that will help them get to the core of how and why they do something. For most, identity is a filter that is so removed from conscious awareness that most will report a completely different perspective than the one their actions reflect. The "say-do" gap that emerges as part of our exploration of users and their lives is usually where the most relevant insights come from. For example, when researching user's perspectives on online identity, most reported that they prioritize security above all else. When asked to show a banking app on their phone, most were found to not use the security pin to lock their phone. Why? Because they found it more convenient to use the phone that way, but they still viewed themselves as someone who values security.

Dig deep and you find the gold. Look at habits. Look at mechanisms. Identify patterns of similarity and difference. Figure out what drives us as users. Design and test. Then optimize. *Design for* growth and change.

We all have the potential to be designers (even if it's only of our own lives). The next chapter provides an overview of the Design Thinking approach and its application to digital design, in order to help us get to know our users, ourselves, and our context better, and design more fitting and relevant products in times of rapid change.

Design Thinking in Action

An Approach for Better Design

"When you start to develop your powers of empathy and imagination, the whole world opens up to you."

—Susan Sarandon, actress

There are infinite ways to create something and infinite ways to imagine a future where the designed product and/or service belongs. How might someone perceive this product or service? How might they engage with it? How might it shape the disruption and/or formation of their identity? There are more questions than we have answers as we set out on the journey of digital creation. What we can say for certain is that the world around us is rapidly changing and that we have little left to do but to adapt to the rate of change in a strategic way. What this means for the digital designer is to design better, quicker, more effectively, and with more purpose. How? The approach to the design process itself may be that strategic advantage.

In a world of constant change, we cannot reply on assumptions generated from past events, states, or perspectives. This is why the approach we take to create products and services needs to accommodate our changing reality. Our reality is composed of not only what we have and what we do but also

© Anastasia Utesheva 2020
A. Utesheva, *Designing Products for Evolving Digital Users*,
https://doi.org/10.1007/978-1-4842-6379-2_4

how we see ourselves and our role in creating our future. This means that when we design, we need to design without prior assumptions and treat each project as a fresh chance to get to know our users and where they are in their journey. It may be true that there is no such thing as the "right way" to design. Design often happens in a messy, organic, and often unpredictable way. However, there are approaches that are more suitable to times of constant change than others. One of those approaches is Design Thinking.

Design Thinking is an approach that works on the basis of rapid evolution of ideas and evidence-based decisions. It is an approach that enables us to adopt the role of curious researchers (as well as strategic designers), and this opens the world up to us and our imaginative capacities. We are free to drop any notion of knowing what's best for the users or the situation and go on a curioisty-filled journey of discovery, exploration, and experimentation. The core value of the approach is the focus on discovery and the constant evaluation of phenomena from multiple perspectives and using a variety of lenses. We form our comprehension from the examination of multiple perspectives of those that constitute or are affected by what we are examining and/or designing. The formation of our perspective in this manner gives us the capacity to deeply comprehend the phenomenon and those affected and meaningfully design for change from the start.

Design Thinking also gives us the opportunity to co-create with those we ordinarily might not even get a chance to talk with, let alone collaborate. It gives us the opportunity to think differently, think in ways that seem silly at first but actually give us a true glimpse into the core of the phenomena we are examining and designing for. Design Thinking is an approach that enables multidisciplinary teams to work together to produce results in a structured, yet very flexible way. It allows the project to meet budget and timeline constraints but gives the team enough room to explore, empathize, try, learn, and iterate on their ideas throughout the process. In this way, the approach allows for unstructured or semi-structured exploration, while the overall process is structured enough to work with the very real constraints of a commercial environment.

Out of the plethora of project management, software development, stakeholder engagement, and general management methodologies in practice, the best for the research and design components in the digital realm is Design Thinking.

What Is Design Thinking?

Design Thinking is a type of human-centered design, which originated from individuals with an industrial design background applying their skills in the commercial world of 20th-century consumerist culture. The core of the approach is putting those affected at the center of the design challenge, such that every decision is framed around their needs and preferences.

The approach provides several advantages over other popular methodologies in that it allows for trial, iteration, exploration, and rapid prototyping of ideas. It also emphasizes the need for the design team to get to know their customers/users and to empathize and comprehend their reality as holistically as possible in order to come up with designs that truly resonate and add value. In a lot of ways, this approach proved much more suitable to not only design of material objects but to experiences, organizational models, culture, strategy, and other intangible elements of organizations and commercial phenomena.

Design Thinking differs from other methodologies in that it:

- Is iterative and nonlinear

- Enables early prototyping

- Requires a multidisciplinary team of stakeholders to genuinely collaborate

- Allows for a variety of voices to be included in design

- Relies on data collected from users from the onset

- Is based on evolutionary assumptions and enables design for change

Over the last four decades, the approach has been pioneered by organizations such as IDEO in application to solving complex global challenges (a.k.a. "wicked challenges"). These challenges often relate to marginalized or vulnerable segments of society and mechanisms at play that are deeply entrenched in evolving cultural practices and historicity of the locale. In order to be able to design solutions that actually work, Design Thinking has been more and more widely adopted by large organizations that are attempting to innovate and provide higher-quality products and services to their customers and communities. The main advantage of adopting Design Thinking in large organizations is that it allows for the team to break out of organizational dynamics (that are often linear and risk averse) that lead to long project and incremental change and be able to come up with ideas for products and services that are evidence based and take exponential leaps in thought. It has also been proven as extremely effective in a variety of contexts, from corporate America to rural Ethiopia, and everything in between.

Naturally, the approach has evolved over time through its application. The approach was first relatively unknown and pioneered by a few organizations actively practicing and evolving the methodology. When the approach came further into the public domain, guides and techniques came to the fore that effectively defined the approach for a commercial context. Generally, Design Thinking follows the diverge/converge process. The team starts with a focal point and goes wide in exploration and abstraction (i.e., diverges) and then through analysis and synthesis comes to a set of insights and/or conclusions (i.e., converges). Typically, this happens multiple times (i.e., iteratively) in a single project. A typical project starts with *discovery* (a process of exploration of the design challenge and its context), then goes into *ideation* (a process of coming up with ideas for possible solutions, which includes rapid prototyping), and *making and taking to market.*

Design Thinking came be known for a common set of methods, which have become a sort of industry standard. These methods include journey mapping, analogous inspiration, 5 WHYs, card sorting, interviews, observations, secondary research, persona mapping, matrix analysis, thematic analysis, ideation, workshops, "how might we" questions, role play, storyboards, data/ process flows, and prototyping. The methodology has popularized the use of post-it notes, sharpies, and whiteboards. It also has enabled project documentation to go from procedural/descriptive large-scale Word and Excel documents to more simple visual ways to represent product/service design (especially through artifacts and showcases).

There is now a dedicated global community championing the methodology and a range of organizations (including Federal Government Departments in multiple countries) that have adopted the methodology with overwhelming success. With more adoption of Design Thinking, the better quality products and services on the market, and the quicker the evolution of users in their habits and expectations. In order to adopt Design Thinking in the earnest, there are a few basic skills that one can refine and apply to solving all kinds of design challenges.

To help digital designers develop the core skills to master Design Thinking, we need to add a few skills to the templates and activities popularized as core to Design Thinking in industry. To this aim, the next section outlines a high-level Design Thinking Toolkit.

Design Thinking Toolkit

To be designer requires a specific skillset. All forms of design involve recombination of ideas and patterns into a new form. But how does one know which form to create? To start the process of discovery (which ultimately leads us to ideation), we need to look at the reality of the phenomenon as it is, not as we believe it is or wish it to be. Once we do that, we are then in a position to transform what is to what we want it to be. To do that well, we need to master a few basic skillsets. These skillsets are the foundation for the Design Thinking Toolkit.

The toolkit is composed of what has colloquially been termed as "soft skills"— skills on how to be a good listener, how to put information together well, how to represent and communicate information, and how to navigate the unchartered waters from exploration, experimentation, and discovery. This is not by any means a definitive set; however, it may be of use as a basis for where to start and what to focus on while on the Design Thinking journey.

The skills in the toolkit are as follows:

- Empathy and curiosity
- Research
- Data analysis
- Data visualization
- Nonlinearity (and intuition)
- Pragmatic optimism
- Inclusive and generative facilitation
- Multimedia communication

Let's go through each one.

Empathy and Curiosity

The first and most important skill is emotive. It is empathy. Empathy is the ability to comprehend and share the feelings of another. It is something that is impossible to design well without. When we design something, we need to take into account the subjective experience of everyone affected by what we are creating. To do that requires us to share the reality of someone or something else, to develop as rich and complete a comprehension as we can, and to share in the experience or emotional reality. Empathy is closely linked to the phenomenon of intuition: when one feels that something may be a certain way but has not yet found a logical/mind way to get to the realization the feeling implies.

The assumption is that everyone is capable of empathy, yet some are more practiced in it than others. There are ways to improve empathy, through careful and directed attention and focus. When one directs their attention to something, they reframe the entire phenomenon through that lens of focus. When one switches attention to another element of the phenomenon (through another focal lens), the phenomenon becomes reframed again in a different way. The insights that come from each perspective form the foundation for recording, analysis, and further insights and conclusions. Empathy here is a powerful focusing lens as it brings to attention the often overlooked parts of a phenomenon, such as emotion and the embodied feelings of those constituting the phenomenon. Insights derived from empathy are best used in conjunction with insights from spoken words and observed behaviors.

When one looks at reality with empathy and curiosity, it is a powerful combination. Curiosity is something to cultivate. It is the capacity to find something interesting or ambiguous, and follow that path until a satisfactory resolution is reached. The main question or filter to view a phenomenon through is "why?". Combined with empathy, curiosity forms a powerful tool for research and design in general. When one asks themselves the question "what does this mean?", "why is it this way and not another way?", "why is it the same/different?", and "how does it work?", we start to get to the mechanisms and core drivers of the phenomenon. Often the paths of curiosity, fueled by empathy, lead to the most relevant and powerful insights.

Don't ignore empathy and curiosity in design. Use them to their full potential.

Research

To be able to effectively collect data, analyze data, generate insights, and improve designs, a digital designer needs to be a researcher. Primarily, a qualitative and mixed methods researcher. Why? Because when we design digital, we design an experience and experiences are complex holistic phenomena that are constituted by evolving interactions/input from both the user and the designed product/service. And we cannot get to this with numbers or inferences about behaviors of populations.

After the project brief/scope has been confirmed, the first step in the Design Thinking process is to create a research protocol that outlines the phenomenon of interest, who the participants are, what methods we use in research, how will we conduct the research, and how will we analyze the data. This document is important as it allows for other researchers to be able to re-create the research at a later date (e.g., with another demographic, in another context, at a later time) and be able to meaningfully compare results. The research component of the Design Thinking process typically involves a secondary research component and a primary research component. The secondary research component allows for the first round of insights and the context/landscape to be established, while the primary research typically dives deep into an aspect of the phenomenon or experience of different user groups.

The research skills of a designer are crucial as they directly impact the quality of data and insights that subsequent parts of the project rely on for as a foundation.

Qualitative, Quantitative, Mixed Methods

There are three types of methods in research: qualitative, quantitative, and mixed methods. The main type for Design Thinking is qualitative because we need to gain a deeper comprehension of humans, systems, and processes, and to do that we often need to either experience the phenomenon or ask a group of humans who have experienced it or are experts in it. The data that we collect in the process is composed of statements, insights, observations, notes, drawings, feelings, occurrences, artifacts, cognitive models, and other things that we come across in our research. We need to be explorative and allow for nonlinear exploration and explication.

Qualitative research is perfect for this as it is primarily exploratory research. It is an approach that is common in social science disciplines and mostly used to explore social relations, including systems and processes with the focal point being the way that humans (and other actors) behave and go about their day-to-day habits and processes. Examples of qualitative research methodology include interviews, ethnography, observations, journaling, narrative analysis, and others. When we engage in qualitative research, we are looking at underlying motivations, perspectives, opinions, and reasons for something happening. From observations and other data points collected, we then generate insights into the problem or develop ideas or hypotheses. Qualitative research seeks to answer questions about *why* and *how* beings behave in the way that they do. Qualitative research in the digital space is often supplemented by quantitative data as well (e.g., analytics or survey data), in order to provide both a population and individual view of relevant aspects of the phenomenon.

Quantitative research deals primarily with numerical data, large datasets, and data models. It reveals a different view of phenomena, one which is based on patterns and trends in populations. Typically, researchers collect numerical data and analyze the data using mathematically based methods (e.g., surveys, statistics). This type of data is data about the quantity of something (e.g., how many, how much, how often, etc.). An example of quantitative data in the digital space is user analytics, which allows us to track actual user behavior, such as click rates, time spent interacting with the product or service, pathway analysis, bounce rates, downloads, shares, likes, and other metrics. Quantitative data provides a foundation for insights into patterns of behavior within and across user groups. It is valuable to use this type of data for A/B testing and deciding which variation of the product or service works best for the majority of users (regardless of their individual experiences/journey).

Rarely do we start with quantitative research in Design Thinking projects; however, often we use it to supplement the insights generated through the qualitative component. This type of research is called mixed methods. It is often used when a digital product or service exists with (an analytics) dataset available to inform (re)design. However, quantitative data alone is rarely enough to develop a rich set of insights as it (by its very nature) overlooks the experience of the individual to focus on observing patterns in the collective. In Design Thinking, we typically start with the perspective of individuals and then look at behavior patterns of the masses, rather than simply look at changes in population behaviors and make decisions based on that information alone. The reason is simple: rarely do pattern changes in the population get you close to the *why* (though they are useful to observe broader patterns of change in the *how* and *what*).

Primary and Secondary Research

Primary research is a must in every Design Thinking project. Primary research is the type of research where you actually go out and interact with a participant, process, or phenomenon. This type of research usually takes longer than secondary research, as the researcher needs to spend sufficient time immersed in the phenomenon to be able to accurately comprehend and represent it. For example, to go into a remote community and live there on-site is far more intensive for the researcher than creating a survey, sending it out to a community, and collecting the results. It also takes a lot longer to interview a sample of individuals from a community than to read interviews conducted by others. The value of conducting the interviews yourself is the opportunity to follow the conversation and delve into aspects that may be interesting at the time.

Primary research stands in contrast to secondary research, which relies on the accounts of others for data points. The difference is you going out to interview a participant vs. reading a report on the topic/demographic published by someone else. Secondary research is vital to set the scene and inform the design of the primary research component. By examining everything relevant that is out there on organizational hard drives or in the public domain, we as researchers start to see patterns emerging that give us the outlines of the landscape that we are looking at. The insights generated through this research form the foundation for shaping our focusing lens and informing the questions and the direction of further inquiry.

Structuring Research

A typical 8-week Design Thinking project includes at least 2–3 weeks of research and 2 weeks of data analysis. These often overlap as the data analysis process starts as soon as we start to collect data. In an 8-week project, leave half a week for the research protocol, 1 week for secondary research and primary research preparation (i.e., participant recruitment, interview scheduling, etc.), 1 week for secondary research, 2 weeks for primary research, 2 weeks for data analysis, 1 week for ideation and prototyping, and 1 week for presentation of ideas. As these activities overlap and happen recursively, these allocations work well for an 8-week project. In longer projects, the rule of thumb is to double the time (for complex phenomena) or run two 8-week iterations. In Design Thinking, frequent iteration often produces higher quality results than spending longer on any particular part of the process. In order to make this timeline realistic, the design team needs to appropriately scope the research and research focus, usually through the use of an overarching "how might we" question and/or a succinct research objective.

Depending on the project, the number of participants can vary from 6 to 12 to 100. The rule of thumb for interviews is around 30 participants (5–6 per user group) or around 30 participants per user group. The reason is that we need to collect the perspectives of at least 5 humans to be able to extract patterns from the data. The more interviews conducted, the richer the dataset. However, it is important to note that for some research objectives less than 12 participants work well, such as in user testing of product designs, where the general principle is "you don't need 100 people to tell you there is a pothole in the road, you only need a few." The intent of selecting participants is to include the perspective of every group affected and include as many different voices/perspectives as possible. A well-designed research component makes the process much easier when extracting insights and analyzing data with the intent to inform design decisions.

Data Analysis

To extract value and make use of the data collected, designers need to be able to meaningfully and rapidly make sense of the data and convert that data into meaningful and actionable insights. Design Thinking borrows heavily from the qualitative data analysis tradition that uses thematic analysis as the primary way to make sense of data and generate insights. Thematic analysis involves grouping and abstracting data points by types, through similarity/difference. In Design Thinking, this process is iterative and ongoing and starts as soon as any information about the phenomenon is encountered.

The process begins by collecting the data and segmenting the data into data points/segments (e.g., quotations, notes, etc.). Then, typically, all the data points collected are put in one big pile and then coded (grouped by similarity/difference). At this level, the insights are grouped by literal similarity (rather than more abstract groupings such as categories and themes). For example, when analyzing interviews of customer service experiences in an organization, the quotations (and other data points) from the interviews are grouped based on the topic or perspective (e.g., like vs. dislike).

Then the team looks at the codes that emerge and generates another round of insights. These insights are added to the other data points, though they can emerge at any time throughout the process. Then the codes undergo the same process as the initial data points, except that the codes are grouped/transformed into a new set through abstraction. The abstractions are called "categories." For example, if two codes emerge "useful contact feature" and "online contact," these would be abstracted into one grouping/category of "contact." The raw data points then also come under this grouping through their code level groupings. Insights are again generated and added to the dataset. Finally, the categories undergo another round of groupings through abstraction to form one to five high-level themes. These themes (and insights generated during the analysis process) form the basis for design direction and decisions.

There are many ways to conduct quantitative data analysis. The type selected depends on the project and data available. Survey data is usually used to generate patterns, disprove assumptions, and to reach more users than qualitative data alone. Usually this is done to provide context, map populations, and generate further insights for design.

Data Visualization

Visualizing data and seeing patterns in data is a core skill in Design Thinking. The general approach to data visualization is that it needs to make sense without explanation and to be easily digestible (<2 minutes of engagement). The value of Design Thinking is in relying heavily on multimedia communication of information such that the team creates more meaning through the representation and avoids creating novels of reports and documents that become too large to be of any practical use to future teams and design initiatives. To visualize something complex simply also adds value through data layering. Data layering occurs when multiple lenses are overlapped in one representation to give more depth and meaning than each dataset provides in isolation.

A common visualization of data in Design Thinking is the journey map, where multiple layers of information are woven together into a coherent mapping of an experience over time. Usually the mapping is **[Journey step]** by **[Action]**, **[Pain point]**, **[Satisfaction]**, and other groupings such as **[Blockers]**, **[Artifacts]**, **[Value Add]**, etc. To be able to create a journey map that captures the phenomenon in sufficient detail, an array of input goes into creating it, such as interview material, documents uncovered in secondary research, observations, notes, insights, and anything else the design team came across in research. The value of the journey map is in capturing the experience in a (relatively) linear fashion in the detail of the steps of the journey, the touch points between different actors, and the areas that work well and those that don't. At a glance, the team can see where/how the design needs to be improved and what/how users are affected most.

Another common visualization, similar to the journey map, is the narrative or storyboard. This type of data visualization may even go so far as to manifest through role play or simulated prototype for a more nuanced representation of the phenomenon, problem, and/or design concept. The storyboard or narrative provides the detail of someone's experience or need through a (relatively) linear step-by-step visual guide. The narrative or storyboard is more emotively focused than the journey map, as it intends to capture the more elusive emotive experience points of a user, rather than comparing information about each point of the experience. The two can work well in conjunction to provide a more nuanced representation of the experience. Both of these types of visualizations can also be transformed into a simplified process map, or call script, or induction process. Anything that is an experience

can be mapped as a process and journey and narrative (depending on focus and level of detail).

Other visualizations include themes mapping, process mapping, graphs depicting statistics and trends, persona mapping, matrix visualization (e.g., difficulty vs. impact matrix), say-do gap visualization, extremes vs. mainstreams mapping, diagrams, drawings, and more. The type of visualization useful for the project may vary, depending on what elements of the phenomenon are important to consider in design.

Nonlinearity (and Intuition)

To do abstraction well, we need to think nonlinearly (and intuitively). When we think linearly, we follow a string of logic that generates itself based on direct logical relations of concepts. Another approach is nonlinear thinking which is a creative, intuitive, and emotional thinking style. The exact algorithms for linear thought are much easier to map and explicate rather than nonlinear thought processes (which can be a bit of a black box to both internal and external observers). However, when one learns to use this type of thought process, patterns emerge that provide value in analysis and design.

When we think nonlinearly, we expand our thought to multiple dimensions and directions, rather than one. We think in mind maps and tree structures, rather than linear one-directional logic. To do so, we often conduct levels of abstraction in our minds on the fly, to see if there are patterns that we can see that spark ideas that we can recombine into meaningful insights. We need to be comfortable to go backward, as well as forward. Look on the surface, as well as examine the mechanisms at the core. We need to be able to see the bird's-eye view of the phenomenon and accurately represent the individual's subjective experience.

When we look at phenomen a nonlinearly, we start to collect layers of perception that create a pastiche through which we can see meaning and generate insights to create more nuanced meaning.

Analogous inspiration is key here. When we look at something through the lens of another (often already existing but different in substance) pattern, we often can learn from what works and what doesn't in another context and apply the learning to the design. It is something that takes practice to do meaningful and effectively, however is a valuable way to conceptualize scenarios.

Designers also need to be able to project manage an iterative nonlinear process of exploration and discovery, which is something that requires not only careful and realistic planning, but an intuitive grasp on how well the team is progressing and when it is appropriate to move onto the next stage of the project. Time × Effort = Result. Time is important to use as a rhythm, as often the number of teams grow in shorter time constraints. For example, if we

need to conduct 30 interviews with a team of 4, then we can realistically do that in 5 days (given that two teams of two researchers can complete roughly three one-hour interviews a day). If we need to increase the number of interviews, we would need to add more time to the project or increase the number of the team. With practice and experience, Design Thinking lead will know when to add time and when to shorten a segment, as needed. They will also know when to do an extra round of analysis and/or ideation and when the team has reached the point of theoretical saturation.

Pragmatic Optimism

To design well requires a certain optimism and a belief that it is possible to achieve what you set out to, using this methodology, and within the constraints of the project (e.g., time, budget, team, access to resources, nature of the design challenge, etc.).

Experience creates the basis for pragmatic optimism as it is often the design lead who acts as the optimism generator through the role of the orchestrator of the project: setting timelines, keeping up with the progress of the whole team, and making sure that the team is performing as a cohesive unit (with no redundant efforts and tasks). The design lead needs to generate energy and keep the team motivated to keep going in the toughest parts of the process (e.g., data analysis and design ideation components). When the team is optimistic, somehow, everything is easier to do and yields higher-quality results.

To be optimistic alone is not enough, as the design leads also need to be realistic in expectations and timelines as the key to maintaining optimism throughout the project is to achieve small milestones continuously and break up more challenging tasks into subtasks that make the process manageable.

Inclusive and Generative Facilitation

To design well is to create a safe and encouraging space for ideas to thrive. To do so, we often need to set a standard for behavior and intent at the start of the project in order to help the team to transition from unhelpful habits to those conducive to thriving ideas. For example, if a team has come from an organizational reality where decisions were typically made by management behind closed doors and the teams were expected to execute upon them without question, then it takes a bit of time to rehabilitate those teams to be fully functional in a Design Thinking project.

Design Thinking in a lot of ways is the opposite of most 20th-century ideals of management and corporate orchestration. Design Thinking was founded on the concept that the idea needs to be nurtured and evolved by a team of multidisciplinary experts (each in their own area) that come together to

create something greater than each of them is able to in isolation. It requires the team to be comfortable challenging ideas, coming up with wild or unorthodox ideas, and not being afraid to go outside their comfort zone to test a concept. They also need to be fully comfortable to champion the voice of the users and stand up for what is right for them. A team that is comfortable to do all that typically works through inclusive and generative facilitation. The team lead sets the tone and the team each assumes the responsibility to being inclusive of all voices and generate participation from all members (including users) in design.

The guiding principles for this type of facilitation are as follows:

- All ideas are welcome.
- There is no such thing as a bad idea.
- Co-creation is key.
- Include all voices/perspectives.
- Look for similarity/difference.
- Creative tension is good.
- Focus on the process, not the outcome.

As with any endeavor, the best way to lead is through example. With teams that are new to the approach, it helps to draw out the first brainstorming session on the board to provide a visual aid for the team to see how ideas form and how to create a useful mind map going forward.

An effective way to do so is through the use of post-it notes. Let's use the example of a values co-creation activity. Get the group to come up with five values that they believe are important for the team to embody. One value per post-it note. Give them 2 minutes to complete the activity. Then ask the team to play a game of snap: someone volunteers a value and anyone who has anything similar puts their post-it note up as well. Collect the post-it notes and group on the wall or whiteboard. Add a post-it note summarizing the group to the center. Continue with this task until there are no more value post-it notes left among the team. Then synthesize the values to a set that captures each value clearest. This activity should take no more than 30 minutes and allows for each participant to be heard and have their ideas contribute to the outcome equally.

Part of the role of designers is to be able to be an active participant in each activity. That requires advanced listening skills, mature emotional skillset, selfless participation, and the ability to lead and follow when appropriate. It also requires the individual to become a selfless conduit for ideas and to allow the idea to form without intervening with personal bias to trying to sway the outcome in a certain direction. This is different to making design decisions.

Multimedia Communication

As part of the Design Thinking process, ideas transform through multiple media and, as such, the designer needs to be well versed in transformation of thought across a variety of instantiated patterns. What this means in practice is that a designer needs to be comfortable composing a concept through whiteboard mapping, walls of post-it notes, prototypes, documents, drawings, and other multimedia representations. Without this core skillset, a designer may overlook an opportunity to represent something more effectively or create the conditions for an epiphany to occur through analogous inspiration. Learn to represent ideas in multiple ways and through a variety of media, and you will find the design process much easier and more fun to do.

The meta-skill for effective design is to *know yourself*. A complementary skillset to this is *thinking*. Thinking is a skill, even if one that we are not explicitly taught in modern educational paradigms. To think is to be able to isolate information, analyze information, put information together in a meaningful way, and draw meaningful conclusions. There are many types of thinking patterns, and all depend on the schemata that you have within your cognitive systems. Thinking is also a learned skill that you can improve over time. One of the best ways to test your own thinking patterns is to sit down in a quiet setting and try to NOT think. The response of your mind and what it brings to your conscious awareness can be very revealing. Another way to test the way you think is to write down a set of statements that outline how you came to a conclusion or a decision about something. Examining and reflecting on the process can be very revealing in terms of how you see the world and make decisions more broadly. If done with honesty and open curiosity, it is an invaluable activity that allows for deeper self-awareness and, ultimately, empowerment.

As a designer, if you are not fully aware of your own preferences, skills, agendas, and biases, you may be inadvertently trying to skew the design to something that you value, rather than something that actually brings value to real-life users. In order to truly know yourself, it helps to be aware of how you think and how you put information together. Especially important is uncovering the assumptions that you (and your team) bring to the table and to test or challenge those assumptions at the start of the design. One of the biggest issues in ineffective or convoluted design is often misaligned assumptions. When assumptions are known, tested, and realigned as part of the research process, the design stands a much better chance to hit the mark when released into the wild.

On a final note on skills, let's not forget creativity. Put simply: creativity is not really a skill. Creativity is an outcome of the process. Refine and practice the skills outlined above and the designs you and your teams create will be much more "creative" and innovative. Real insights come from real humans. Ask them and you'll be amazed what you find.

Design for Real Needs of Real Humans

The reason that Design Thinking is arguably a better approach to design than others is that it accounts for the real needs of real humans (rather than an abstract concept of "user" circulating the project room, or the assumptions the designers may have about someone they know nothing about). It takes the guess work out of design. Either the designers ask or the participant tells them. This is also why the research and data analysis component of the design toolkit is so important. Without good research and insights generated from the research, the quality of the design goes down. With good research and insights, the quality of the research exponentially increases the more the design team is able to learn from the users about the phenomenon, experience, process, or perspective that is inherently not native to them.

The purpose of this chapter is to provide a brief overview of a way for designers to do what they do best, better. The core reason why this approach is arguably better than others is that it bases design on a deep empathetic comprehension of the phenomenon relevant for design and always accounts for the multiplicity of perspectives that our convergent reality constitutes. The way to make sure that we are capturing clear data from the participant is to observe and listen far more than talk, and be very careful to check your own (and the team's) assumptions before doing the research. The reason that Design Thinking is so powerful is because it necessitates asking real humans and listening for real answers. Answers we may not like or expect or comprehend at first, but ones that offer a genuine look into someone else's reality. To do that is simple: listen at least 8 times more than you speak in any kind of research engagement (e.g., interviews, ethnography, workshops, observations, etc.).

For example, in an interview, do not add content into someone's account of themselves. Keep silent, no matter how tempting it is to complete the sentence for the participant, and especially when they are struggling to find the right word and it feels like it is on the tip of your tongue. Resist the urge to speak at all costs, because if you speak up, then effectively you are skewing the interview and rendering the data less valuable than if the participant was left to articulate their thoughts and their thoughts alone.

One of the best tips for beginner researchers is to stay silent when a participant falls silent, and remain silent for at least 10–15 seconds before asking a follow-up question (e.g., "why?" or "how?"). It is harder than it seems. To get better at conducting interviews, try the technique with a friend first. When having a conversation with said friend, wait until they ask you an open-ended question. It needs to be an open-ended question, as yes/no questions are not very useful for thoughtful or extended answers. Generally avoid using closed questions in research (unless asking something very specific). When they ask you the question, pretend to think about the answer while counting to 15 in your head. Notice how many of those you try this with are able to wait you out, and how many jump in and fill the void left by the silence. Observe yourself reacting in the same way next time a friend pauses the conversation while trying to find the right word.

Asking questions at the right moment, however, can be of immense help to the participant: questions such as "why?", "how did that happen?", "how did you feel?", and "tell me more about...." These kinds of questions act like a springboard for ideas, and because they are open ended, they open the conversation up rather than closing it down or skewing the flow.

Note that if you notice the human running the interview to be speaking a lot and finishing the sentences of the participant, then put a mark next to the interview number and use the data with caution. That is how untrained (or biased) researchers skew data to their needs. It is not a helpful practice if you actually want to design for real humans, rather than for what you assume they want and who they are. To reduce your own bias, avoid writing down things you cannot directly observe/experience (unless you clearly mark them as your interpretations).

Also, when starting out on the journey of Design Thinking, be careful of research that you know has been conducted through tailoring of data to make one aspect of the phenomenon seem more important than every other (especially without a good reason). It is common in untrained teams to jump to conclusions too quickly and leave the ideation half baked. The results are usually not terrible but often lack a certain resonance and impact that deeper research enables.

Don't be afraid of the reality of the situation, even if it is uncomfortable. When we look at what is with empathy, openness, and compassion, the world transforms from a place of darkness to a place of kindness and opportunity. Most of those you engage with are more than happy to help. Most also appreciate having their voice heard and acknowledged, and for some their voice is meaningfully heard for the first time. That is a lot of responsibility and that is why, as designers, we come last and the needs/priorities of those we are designing for come first.

Design Thinking for Identity

Where does identity fit into all of this? Identity is core to the human operating system and can be viewed as an interpretation filter for an individual's reality and subjective experiences. But to truly get to know someone, you cannot simply take their account of their identity at face value. Often human beings aspire to be something while at the same time embodying something completely different. That is where the majority of say-do gap findings come into being. These findings are the gold that we seek to find as designers. They give us a glimpse at the truth behind the veils of perception and give us a chance to see motivations and drivers for certain preferences and behaviors.

When someone has a perception of themselves that they are actively trying to cultivate or are unconsciously harboring, then they are more likely to report contradictory priorities, perspectives, and needs to those that they actually practice. That exploration is the bread and butter of insights. If everyone was capable of sophisticated self-reflection and reflexivity and practiced that every day of their conscious lives, then research would be a lot simpler. All we would need is a self-reporting digital tool and we would be able to find most of the information we need for intelligent design decisions remotely and through the aid of sophisticated algorithms. However, that is not the case at the moment and the conscientious researcher needs to carefully unravel the truths of the individual and help them realize for themselves where they actually are positioned, who they are, and why they do what they do. It is an art and science, and requires emotional maturity, intelligence, empathy, and kindness to do well.

Let's return to the reason why we are focusing on identity in this book on digital design. In the previous chapters, we have explored how the modern human came to be the way they are and the different ways that they perceive and navigate our shared reality. When we looked at the evolution of information across time, we observed the trend that in the last 100 years or so, humans have become more focused on cultivating an individual unique identity than (arguably) at any point in the past. The reason may be in part to do with the new global reality that digital media has enabled. Or it may be a reflection of the evolution of our species more broadly as we are now more able than before to assume responsibility and accountability for our lives and the choices that constitute them. In any case, for the first time in history, we as a global collective have the opportunity to reexamine every system that we are part of and that exist, and change the trajectory our species (of all life) in the direction we consciously choose and want to go. At the heart of this transformation is our relationship with and perception of ourselves.

In order to redesign ourselves and our lives strategically, we need to explore the formation of and design for identity. The next chapter provides a detailed exploration of how we can do that in practice.

Design for Change

Creating Digital Products and Services That Enhance Lives

"Once we accept that we are a product of our culture, we can begin the act of deliberately redefining our sense of self."

—Kilroy J. Oldster, Dead Toad Scrolls (2016)

There are two mechanisms (or types of mechanism) that are crucial to digital design: the formation of the individual and collective identity. In the previous chapters, we have explored how identity forms and its ongoing role in our co-created reality. This chapter is dedicated to one of the most important aspects of design, digital or otherwise: monitoring the impact of our designs by tracking and monitoring the waves of change through the filter of identity (both individual and collective).

Why identity? As we explored in the previous chapters, it is the core mechanism for anyone to see and experience reality through. A set of thought patterns, meaning and belief systems, and attributes (and relationships between attributes) that are used to define someone or something. Identity shapes not only behavior but perception. Change perception, and behavior

© Anastasia Utesheva 2020
A. Utesheva, *Designing Products for Evolving Digital Users*,
https://doi.org/10.1007/978-1-4842-6379-2_5

changes (any attempt to do it the other way around leads to temporary results at best). Change perception and the very nature of reality changes, along with everything in it. It is also the clearest way to explore and comprehend paradigms, the biggest levers we have for change.

As we explored in Chapter 1, the nature of human reality is constituted by what we can call "immaterial" abstract ideas and concepts. These concepts emerged from representational systems that have evolved us and through us, and have over time created a dataset that each human inherits as part of their assimilation and acculturation to their surrounding context. Historically, these representational systems have been founded on the idea that we can use a symbol (or symbol set) to represent, communicate, and make sense of what we experience in our external reality (typically external to the physical body, but over time becoming external to our sense of self). These symbols were static, as were the associations between symbols that were instrumental to creation of meaning. As the symbols and relationships between them were static, the meaning was commonly assumed to be static too. These meaning systems over time evolved into what we commonly call "culture."

As culture evolved over time, throughout disparate locales, culture-specific similarities and differences emerged. Culture was inherited and iteratively enacted to shape subsequent generations, and through us evolved to its current form. Different cultures have had different dynamics between the individual and collective identity. The most recent variation of culture, in our global melting pot of everything that came before, brought into focus the importance of alignment between individual and collective identity. We arguably now place far more emphasis of being an individual than we have in previous generations or in family or community settings. The individual was a layer on top of the collective identity: the collective identity held the axioms and the individual was afforded slight variations that were mapped to a unique identifier (e.g., the name on your birth certificate). The 20th century was the first time in recorded history that a monumental cultural shift occurred, such that the equation was flipped on its head. We are now individuals who are part of a collective called "human." Rarely is it still in good taste to blindly adopt a collective identity and its set of axioms as the basis for who we are. Especially a legacy one (i.e., one that came from the past). The individual identity is now the focus, and this identity can be created from literally anything someone can be exposed to (with digital media, that is an increasingly large volume of increasingly diverse information).

Where does that leave us? Living in the 21st century with a smartphone in one hand, a laptop under one arm, racing around trying to find purpose and meaning in a brand new reality where what came before no longer matters as much as what comes next. We are in the most vulnerable position and, simultaneously, in the most powerful position to attain true freedom (i.e., the freedom to do and be whatever we want to be). By redefining ourselves, our role and purpose, and aligning the individual and collective identity through what we want our reality to be, rather than out of obligation to perpetuate what came before. Is that a bold idea? Yes! Is it *the key* to our individual and collective metamorphosis? Yes, with the right perspective. Because we live in an era where a shift in paradigm can be felt in less than 12 months, and cemented in less than 5 years, a lot can change just by focusing on changing our minds. Is digital our best friend in our journey of metamorphosis? Absolutely, because it is the clearest reflection and extension of what drives us: our paradigms (specifically the paradigms we use to define ourselves).

Use Perspectives to Strategically Improve Quality of Life

Perspectives are incredibly powerful, but they are often taken for granted, or placed in the background, or downright ignored. Yet, there are times where their importance becomes impossible to ignore. Every once in a while, we have that moment of perfect clarity. A moment where we realize all that we are and all that we can be in a way that stops us in our tracks and forces us to reevaluate everything we thought we knew. A moment where we see the ongoing process of how we are constructed and how we (re)construct ourselves. That moment is one of the most valuable that we can experience in our journey of becoming. Becoming what? Well, that's up to each and every one of us to decide, on an individual and a collective level.

That moment of clarity is the epiphany that enables us to pivot and change the course of our lives. Where the veil of "normal" (i.e., unconscious) is lifted and we are forced to confront ourselves as we perceive ourselves and/or as we are perceived by others. It may be a moment when choosing a restaurant and realizing that the ones in your area are so perfectly tailored to your demographic and financial position that you really only have the same two options you usually go to. Or picking out an outfit for an event, and realizing that the layers of fabric you wrap around yourself with are nothing more than smoke and mirrors for an image of someone you need to pretend to be for the occasion. Or a moment where you look at your social media feed and realize that not one person on there would help you move house or care if you were having a bad day. Any event that triggers the reevaluation of the core of someone's perception (and thereby their whole being) by putting information into focus that does not fit the usual operational paradigm of the individual. Use that moment as a start of an exploration of your own paradigms and what thought patterns create/enact them.

That point of self-awareness is one of the most valuable experiences that one can go through as it allows for one to see beyond the facade of identity that one has spent a lifetime carefully cultivating and get a glimpse into the core of who and what they truly are. Once the terror and discomfort fades, that is when the true opportunity for growth emerges. An opportunity to question everything that came before and pick and choose what we take into the future. A way to evaluate the individual and collective identities that one embodies, debug and optimize, and enhance what works.

How do we do it? The best place to start is through the digital media that you have within your reach. Take your mobile phone, for example. It likely has a host of apps on it that you use to navigate your reality. A few of them are extensions of your memory (e.g., Photo Gallery, Dropbox, Calendar, etc.), your awareness or knowledge (e.g., Fitbit app, Google Search Engine, etc.), your community or social reach (e.g., Twitter, Facebook, etc.), your skills and reflexes (e.g., racing or arcade-style games), and your capacity to represent and communicate (e.g., email, chat apps, camera, Instagram, etc.). Just by looking at the types of apps that you use most frequently and mapping your use patterns throughout the day can be a revealing activity, where you see yourself from the perspective of investment of time into product/service/activity. From there you can begin to see the patterns of action that constitute your day-to-day.

Diving deeper, we can look at the *why*. Why are we using this app over another? Why are we spending 60% of our digital activity on gaming this week when we dedicated 80% of our time on social media last week? Why did we delete that app only to install it again a month later? Each of these patterns points us to the decision that we made in order to enact the pattern. That decision is the most important piece of the puzzle. The *why* that drove our behavior. If we were to look deeper, we would look to the content itself. A highly useful activity is to map the type of content that was sought, engaged with, or created, in order to begin to paint a more complete picture of decisions and the context for those decisions.

Simple activity to get us started: go back to the first post or picture of yourself on Facebook (or LinkedIn, or any other digital chronology you have created for yourself over a period of at least 2–6 months). Look at every activity since that first post and begin to map your portrayal of your identity. Keep in mind that the record is a public front of your identity that you have carefully cultivated and that there is infinitely more left out than represented. It is akin to a showroom. A showroom may not give us any insight into the full supply chain or the complete set of the thoughts and feelings of the board of directors, but it is enough to anchor our inquiry. When conducting an inquiry into the self, start with the obvious (i.e., what you can observe and/or experience) and look beyond it into the mechanisms that generated it. If you do this process in the spirit of open inquiry and curiosity, you may find yourself with a picture of your identity that you did not expect. That is the "Aha" moment that is the trigger for the journey to carefully changing aspects of your reality to more fully align with what you intend, want, and wish to embody and experience.

Digital is the best tool for this process, as it is the medium that we engage with most intimately. Yet, as digital designers, we often overlook the full extent of how a digital product/service fits into a user's life and how it may shape them, their reality and operating paradigms over time. The phenomenon of the "Like" button is yet to be fully comprehended in terms of the social changes that it enabled and drove. How might we have behaved differently on Facebook or Twitter if that feature was never released? How might we have shaped our reactions if we had more options available from the beginning, rather than just "Like"? That is just one feature. Digital design is a lot of responsibility.

If we design from anything other than a genuine intent to improve quality of life and provide more freedom to the individual, then we run the serious risk of skewing users into trajectories that are less than ideal (e.g., freemium apps that have paid features that prey on those with addictive tendencies). If we design an app that deliberately limits freedom (e.g., mass monitoring apps), what are the ethical implications of what we are creating? If we design an app that exploits or emphasizes the worst qualities in us, then what impact are we putting out into the world and what kind of future are we orchestrating into being? We need to think as far into the future as we can when we design something, so that we can mitigate negative impacts and amplify the impacts we want. How might we do that? That is where we come back to the idea of alignment between the collective and the individual identity. When an individual being themselves is benefitting the collective, and the collective benefits from any/all individuals from adopting the practice, that is when we know we are on the right track.

The power of digital is to enable change. One of the primary responsibilities of digital designers is to design for change. To assume that users are static, unchanging entities that we can categorize once (according to whatever criteria we deem fit) and they would remain that way forever, is naive if not plain ludicrous. When we experience our users and, to a greater extent, ourselves evolve through the products we create in a positive way, that is a point of earned pride for a designer. In designing for change, for meaningful and purposeful evolution of our users, it is crucial to constantly observe and monitor the impact(s) (individual and collective) of our creations and ensure that what we produce improves life, rather than degrades it.

Evolving Identity Through Digital

In order to be able to strategically evolve, we need to have as nuanced a comprehension of identity as possible. Specifically, we need to have a solid grasp on how the individual and collective identities form and evolve, and the relationship between the two for any user or user group. To learn to do this well, it is best to start with yourself. The reason is simple: you have more access and reach within your own consciousness than within another's. You are also able to compare logic and feeling firsthand, and that will enable you to see patterns that you may otherwise dismiss the importance of or overlook all together.

Let's start by mapping the current state. Remember: a curious and open mindset (without preconceived notions of "good" or "bad") is crucial here. To adopt this mindset it helps to change one word within your cognitive system: "judge" becomes "evaluate." If you cannot judge anything, you cannot get into trouble with preconceived legacy bias. You can evaluate anything to the Nth degree from infinite perspectives and reach a plethora of conclusions, yet none of it will evoke a fear-based response of something being "bad" (especially something beyond day-to-day conscious activity). Instead everything becomes either "useful" or "useless" to the selected reference point. The other reason that the concepts of "good"/"bad" as definites are not useful to inquiry is that they are (a) subjective, (b) absolute, (c) need to be handled with caution and purely from a relational (i.e., in relation to direct context) perspective to navigate without condemning yourself or others (or misinterpreting something vital).

For example, most people you ask will tell you animal cruelty is "bad," and in the same breath admit that their favorite fast-food burger place is "good." Unless you point out the contradiction, it is unlikely that the misalignment will be noticed or felt by that person.

Most of us operate that way: by classifying in relation to what we want or want to be associated with, with limited examination of the alignment or misalignment between what we believe and wholism of reality. So let's start with getting to know ourselves, so that we can learn how to get to know others and design for and monitor the impact(s) of what we create as holistically as possible.

Mapping Identity

Here's a simple exercise for mapping identity:

1. Get a pen and paper out (or post-it notes, if they are at hand).

2. Set a timer for 5 minutes.

3. Take a deep breath and clear your mind.

4. Write down as many sentences that start with "I..." as you can to describe yourself. Don't think about it; just write everything that comes to mind.

5. Once the timer goes off, put the paper aside.

6. Do steps 1–5 again, on a different sheet of paper (or different set of post-it notes).

7. Take the lists of "I" statements and sort them into groups of sentences that start with

   - "I [verb]…"

   - "I am…"

   - "I have…"

   - And any others.

8. In each group, sort the sentences into similar buckets (i.e., subcategories) and then group those buckets into bigger buckets (i.e., categories).

9. Group the buckets (i.e., categories) into groups again (i.e., themes). Do that until you cannot group any more. You should see clear groups and themes emerge. Pay attention to any outliers (i.e., things that don't fit)—these are the most useful to examining the mechanisms and paradigms at the core because they break the dominant perception thereby bringing it to the fore.

10. Review the themes and map it all out on another piece of paper.

The activity will provide you with a range of statements that you can use to derive the paradigms that you operate through, and the values that you embody through your identity and behaviors. If you do this comprehensively enough (i.e., honestly and shamelessly), you will start to see the misalignments and disconnects quite clearly. Once you see the things that don't make sense from the perspective that you are using for analysis, then you can start to delve deeper into *how* and *why*. The process is like pulling a thread: once you start, the whole thing unravels to constitutive pieces and you then have the opportunity to put it back together in a better way.

If a part of you on some level resists this exercise, dive as deep as you can into that rabbit hole and figure out why there is something that you hold within yourself that is actively attempting to prevent you from self-examination and redefinition. Arguably, there is no perspective, perception, thing, or experience that is worth stagnation, or worse, going backward. To know which way is which, the compass of values is incredibly useful.

The next part of the activity is to use the dataset created through the exercise above to derive the values that you embody, and compare and contrast them to the values that you want to embody in the future. Useful questions to ponder are as follows:

- What values can you derive from the "I" statements?
- What values do you want to embody (i.e., what kind of human/person do you want to be)?
- What "I" statements support those values?
- What "I" statements do *not* support those values?

Let's look at the difference between the types of statements. "I…" statements that use a verb will allow you to look at actions and behaviors that matter most to you. "I am…" statements will allow for you to see what static attributes and descriptors you see yourself as embodying most or are important to you. "I have…" statements allow you to see what material objects you use to define yourself. "I will…" or "I would…" statements allow you to explore conditional responses and states. There is a lot that you can learn about yourself (and the users you design for) by exploring and unravelling how you view yourself, and, by extension, what matters most to you.

A note on "I have…" statements: they are a great way to trace what matters to you, but a very limiting way to define yourself. The reason is that that perspective and self-definition is an inherently dangerous one that has gotten us, as a species, into a lot of trouble in the past. Cultural legacy has been largely founded on the concept of "I have" and then "I am" to justify the "I have." "I embody [value]" or "I [value]…" is a much more powerful position than "I have [thing]," as material things are only representations of something and rarely extend to the core of who and what we are. Though at times, we can use material objects around us to map the say-do gap (e.g., someone says they are health-conscious but their fridge is full of processed food). We cannot assume anything about someone's operating paradigm until we map behavior to intent, intent to values, and values to a sense of self that the individual has and/or wants to cultivate. This is where we find great value in adopting a materialistic lens and, at times, relying on the material things around them to, through further inquiry, start to get a glimpse at the core.

On that note, if you find yourself struggling with "I" statements or want to take the activity to the next level, create a fourth list of "I have" statements. For each "I have" statement, add a column that finishes the sentence "and this is defines me because…." The second part of the "I have" statement is what will get you to the themes for analysis. Also worth noting the branding of the product/service and the reasons why you are/were attracted to the object (as well as the strength of the attraction). The way that provides the most personal freedom is when the "I" leads the "I have," rather than the other way around.

The final part of the activity is to map literal behaviors (i.e., thoughts, feelings, emotions, actions, discourse, etc.) to the value and "I" statements. This is where the gold typically hides: in the say-do gap, and the gap between perception and material evidence.

Comprehending Ourselves

Since recorded history, we have come a long way into the quest to comprehend ourselves and what it means to be alive and, by extension, what it means to be alive in our current form. Luckily, the distribution and convergence of major paradigms has provided us with some answers that have meaning in the modern context. For instance, we are now free to adopt a perspective where we *are* god, rather than having to adopt an ancient doctrine of *what others thought god was* at any point in the past. From a cultural perspective, that paradigm shift positions us as a part of a wholism, a continuum of life/consciousness of which our material body is a vehicle or an instantiation rather than the be-all and end-all. It is akin to going from verbal stories of our past inherited from selected members of our community to Wikipedia. The paradigm shift has massive implications for both individual and collective identities, and for shaping our evolutionary trajectory.

The core implication: the quicker we let go of materialism and the superficial, the quicker we transition in focus to what we have come to realize truly matters to us in the modern age (i.e., the self, the character, the embodiment of values that we can be proud to stand behind). However, we also need to recognize that the above is just one paradigm out there in the primordial soup of culture.

Once we have gone through the process and have a set of values that you (or the user or user group you designing for) embody and want to embody, then it becomes a matter of scoping the product/service and designing features that enhance and amplify these values and, in the best case, prevent its opposite. Because more often than not the implications and impact of features are only visible in their full extent after the product gains traction in market. Design for the best case scenario and find as many ways to interpret what is being created from as many perspectives as possible as a way to mitigate risk (another reason why designers from multidisciplinary teams are able to create higher-quality products/services—the variety of perspectives is a practical advantage).

Monitoring and Design for Change

As we are feeling and observing more and more in real time, the world we live in is constantly changing. We are transforming faster and becoming more unified in how we perceive and experience reality. We have become connected through our ideas, our desires for the world, that are becoming more unified around the vision of a global awakening through mature sustainable healthy positive peaceful relations. That has been the ideal for quite some time in the current living generations (ask someone next to you how they feel about world peace and they'll likely tell you it's a great ideal to aspire to). However, the crucial difference that took us to the tipping point is that we have become better in the last two decades in our capacity to genuinely do something about it. We have become aware of the conversations that are going on at all levels of organizations, groups, social networks, and other channels. We can no longer be content with the surface appearance or the facade of something. We want to see the reality of the situation and make informed decisions for ourselves and those we care for.

We now realize that the time has come for us to make a choice. A choice of becoming truly sustainable and evolving gracefully, or the choice of collapsing under the weight of the debt of current systems and practices. To comprehend and consciously evolve in the direction we both individually and collectively want to go, we need to monitor and design for change.

There are infinitely many ways that we can measure change. And infinitely more conclusions we can draw from our observations. One of the most important ways we can monitor for change is to monitor for paradigm shifts. When we monitor paradigm shifts, we naturally diverge from a surface comprehension of a phenomenon to evaluating the core values and mechanisms for its emergence. This is particularly important when designing products that have social value and the ripples of which resonate out to the social dynamics that these products constitute and beyond.

When we design a digital product, we are actively shaping a new reality enabled by technology. Technology that reduces the things human beings are notoriously bad at, maximize the things we are good at and love doing, and help us monitor the impact we have on the world around us. The way to do that is by increasing connectivity and enabling the capture and sharing of more valuable and refined information than we have ever had the opportunity to do before (without compromising basic rights of individuals to their data and their life). To do so, we need to design digital products and services that allow for the capture, analysis, refinement, access to, and visibility of different information streams and the capacity to act on information when appropriate.

There is nothing more dangerous or less useful than an identity that is forged out of building blocks of incompatible (primarily legacy) systems. For example, a homosexual teenager raised in a strict Catholic environment will experience this incongruence between who they want to be and who they have internalized that they need to be in order to be accepted by the community. This applies not only to individual's identity but also to group or organizational identity. How many companies have you encountered that have a list of company values, a strong marketing strategy, and a media presence that *almost* supports those values. Then you meet someone who actually works at that organization and realize that they not only do not embody those values but are diametrically opposed to them: "We care about our customers..." but not enough to provide quality products and support, or "We value..." and then do the opposite. We can avoid such pitfalls by emphasizing the role of identity in research and design. Information is what defined us. Information is what we can use to change us. To transform who and what we are, to who or what we want to be.

Identity has been demonstrated as a powerful driver for behavior. The current economic reality has segmented, divided, and pigeonholed us as consumers into neat buckets that the marketing department knows how to handle. These commercial realities have spawned identities that we may or may not like or want. These types of identities are based on what we can afford. The digital space breaks the constraints of such identities. In the digital realm, one can be anyone and say anything without being constrained by the attributes that have pigeonholed them in the material reality that they constitute. In the digital space, the focus is on ideas, and the formation of identity is much more derived from the value of the content, rather than the package.

In the digital space, it only matters what gender you are if you need to have a real-life picture of yourself (e.g., government identity services). One can always try on a different gender by making up a fake persona and playing out scenarios that they would otherwise not be able to. They can also carefully cultivate their offline identity and extend it by making it more valuable. Social media platforms take that concept to extremes by allowing users to cultivate their personal brand. There is nothing stopping someone from marketing and earning money from their image and what they represent. These types of digital identities entrench the user into a certain identity. Losing one's social media account would be the equivalent of losing your home or your bank account. The value of the asset appreciates over time and it is often these metrics that we reference as giving us value and relevance. Our identity has effectively become an asset. And often it now has little to do with what we do, but what we represent ourselves as. There is a dark side to this phenomenon, where cognitive dissonance can occur if one spends enough time being validated in the digital space vs. in the offline space (especially if there is a grand canyon of difference between the two).

The way to cultivate identity and use it for good (not evil) is through strategic design of the affordances and constraints of the medium such that it validates the parts of the identity that the user wants or needs to cultivate. A common example is a "Like" button. The "Like" button has revolutionized social connection far more than we realize. Now it has become synonymous with attention in general. It has become a metric. A point of validation such that it is often the target of posted content, rather than the substance of the content itself. How many advertisements are there that claim to raise your followers (and their "Likes") by tailoring content to appeal to what we think (or the advertising company thinks) the users want? How many times have we found ourselves looking in vain for depth of content only to find clickbait? How has that changed how we view content, ourselves, and the products and services we engage in?

The value of digital is to accelerate the maturation of our consciousness and sense of self by bringing to light the things that we consider core to our being. They allow us to be free to explore what we want and what we need to become. We can now become someone that is genuinely transhuman (i.e. someone other than the static attribute of "human" implies). We are no longer constrained by biology. It is time to adapt to that and become more than we can conceive of as possible to do that, we need to keep identity flexible and values based, design for change, and monitor the effects of our creations on both our subjective and shared reality.

Disrupt identity for good. When you design a set of attributes that define your users, go beyond the static and connect deeply to the core of who they are and/or aspire to be. Design for change, even if it is as simple as a change in the perspective of yourself.

A potential new paradigm: life ➤ human ➤ [name].

Measuring Impact

When we set out to (re)design our reality or embodied sense of being, we need to figure out how we can evaluate the direction we are actually going, and whether or not our interventions and mechanisms for change are successful. Typically that involves painting the picture of current state, learning from the past, and iterating the future. Along the way, measure strategically in order to grow effectively, through measuring the impact of digital design decisions on identity and paradigms of the user(s).

Digital provides a powerful means to measure both change and impact. Digital does so by providing a way to collect potentially infinite sets of qualitative and quantitate data that we can use to observe patterns. Digital media also has the unique quality of inbuilt self-reflection and reflexivity. It is able to collect data from users and about users/use, and about its own performance. That is a powerful combination if used with purpose. What we measure ultimately becomes the lens through which we view reality.

The primary focus at the start of the 21st century is monitoring for three types of impact: economic, environmental, and social. The emerging focus is on measuring paradigms and their impact on the other three types of impacts. Consider it a meta-driver that shapes the rest. To assess any type of impact, we need to monitor how the product and/or service is changing the current state. What improvements is it creating? What negative effects (if any) is the product and/or service creating? What does the world look like before and after the product?

As digital designers, we also need to be careful to consider the impact of what we are creating and how the product/service may shape the well-being of our users. For example, we may create an app for a marginalized segment of our community with the aim to enable them to feel empowered. To force them to compete and award points based on built-in criteria is arguably not the best way to go. Why not? Because competition has been shown to decrease the feeling of empowerment rather than increase it over time. It is the same as the feeling of intrinsic accomplishment (e.g., finishing a painting) vs. extrinsic/ attributed accomplishment (e.g., winning a contest). If we, on the other hand, create a digital space for individuals to share stories, connect, and collaborate on initiatives that are most relevant and impactful to them, then that creates a whole different dynamic, effects, and social impact (i.e., feeling of having the power to change the situation and the results to prove it). It is important to always keep in mind how the product/service we create will shape not only how someone behaves, but how they perceive themselves through it, and what value they gain through long-term engagement.

A relatable phenomenon for digitally driven phenomena of such change is the Google Search engine. The search engine allowed for the first global social experiment in organizing and finding information generated by an uncontrolled, unpredictable group of users, no two which were alike. It paved the way for our organization of information in our own minds and the eventual adoption of Google Search Engine's functionality to our cognitive abilities and the way that we form schemata. We are now more educated. As popularized by the sitcom *How I Met Your Mother*, a social debate/argument is no longer something that commonly occurs because anyone can at any time search for any piece of information that they want. The word "google" has even become a verb. An action that someone can do that leads to a replicable outcome (within certain constraints). That in itself has transformed culture. To be without Google is almost akin to what being illiterate was 100 years ago. That cultural change is something that we can monitor. The change to our work, our classrooms, our homes, and our own cognitive patterns. We can see it in being more balanced, or more or less risk averse. We can see it in the change that we enact in the world, our priorities, and our topics of conversations online and offline.

One of the most interesting ways to monitor for global change is through tracing the conversations that are happening thematically and conceptually.

We can trace the axioms, ontology, and epistemology of any thought and/or meaning system. Once we do that, we can start to evaluate the readiness of a certain group for specific aspects of change. This way, we can design for changes that are needed and wanted, rather than changes that we incidentally or accidentally create.

The success of the designs we create is arguably in the impact of the design. The more impact, the more change we can observe through the digital product/service and those engaging with and through it. The more desirable the changes, and the more amplified the impact becomes, the more value the product and/or service is able to create. Let's not forget that the picture is only as good as the tools used to create it. The more we triangulate between datasets and perspectives, the more certainly we can claim that our product/service is having a genuine impact.

Digital Trends
Impact of Exponential Digital Progress on Life

"Our intuition about the future is linear. But the reality of information technology is, it's exponential."[1]

—Ray Kurzweil, futurist

In the last chapter, we explored identity, how paradigms shape the formation of identity, and how our identity has been extended through digital. In this chapter, we go one step further and explore how the trends of the 20th and 21st centuries in digital technology have affected our capacity to predict and strategically design and how we might use our newfound freedom to (re) create our identity to transform our world for the better.

Major Leaps

Let's start with a quick overview of the major changes that have transformed us in the last four decades. The major evolutionary leaps in digital technology can be grouped into three key areas: automation, information communication technology, and artificial intelligence.

---

[1]Source: https://www.bizjournals.com/sanjose/news/2016/09/06/exclusive-google-singularity-visionary-ray.html.

© Anastasia Utesheva 2020
A. Utesheva, *Designing Products for Evolving Digital Users*,
https://doi.org/10.1007/978-1-4842-6379-2_6

Automation

Let's look at automation. We have come a long way from the Fordian model of factories that used a strict segmented organization of human labor. To complete the end-to-end production process, humans were given a task and expected to repeat that task over and over countless times a day as part of the production line. These tasks were the kind that humans are arguably the worst at (or at least the least suited to doing for long periods of time): repetitive manual tasks that require no independent thought or creativity. The model had its obvious downfalls and the effect on human cognition was one of deterioration rather than improvement. Luckily, advancements in automation, such as advanced manufacturing machinery, robotics, and specialized software programs (e.g., Enterprise Resource Planning systems), reduced the need for human input and freed human beings to pursue activities arguably more suited to their innate nature (or at least that gave them the opportunity to grow).

The problems with assigning humans manually repetitive tasks include (1) lack of change, (2) lack of learning opportunities, and (3) high rate of human error. Advances in automation resolved all those problems as technology is fantastic for the very reason that humans are not in this context. Technology is able to complete countless repetitive tasks in a fraction of the time it would take for a human to complete and does so with little to no error. The replacement of the need for humans to complete these types of activities opened up the workforce to a new form of much more creative pursuits: those enabled by information technology and, later, information communication technology (ICTs) more specifically.

Information Communication Technology

The most interesting (and arguably the most significant) step in technological evolution thus far is the emergence of information communication technology. The term "information communication technology" refers to a plethora of products that record, store, transfer, transform, refine, and create information. What is "information"? A simple way to view the concept: "data" ➤ "information" ➤ "knowledge" ➤ "intelligence." Data is a symbol or set of symbols that have a specific meaning and maintain the integrity of that meaning in isolation (e.g., temperature, length, weight, an image of a tree, a sentence describing an object). Essentially anything that holds meaning for us (notably, a limited meaning), such as a symbol, can be considered data or a data segment. When data segment (also called data points) are put together with other segments create more meaning in relation to one another than in isolation.

Information is a combination of data points such that more meaning is derived from the set than each piece in isolation. For example, "28°C" is not enough to make much meaning from the data point alone. It could refer to an oven temperature, the weather, body temperature, or anything else. If we combine "28°C" and "Seattle," then the possible information we can derive from the two data points is "the weather in Seattle is 28°C." Knowledge is when two pieces of information are combined to create even more meaning. When we add multiple information statements to the same set, we can create knowledge. For example, if we combine information on the weather in Seattle over the past 20 years, we can draw conclusions about weather patterns which would create more meaning and value than a data point in isolation. Knowledge can be viewed as the creation of more meaning, often more abstracted, that is derived from putting together information in a meaningful way (usually using a compatible type of logic).

Let's take our example a step further and create intelligence. For example, if the weather on October 1 in Seattle for the last 20 years has been "19°C" and for 2020 the data point is "35°C," we can extrapolate that it is an abnormal weather pattern and look for causes for this abnormality. The process and outcomes of the said process can be considered "intelligence." The difference between knowledge and intelligence is the level of meaning and abstraction, and the capacity to make decisions with efficiency and effectiveness.

Please note that there are many definitions of these terms and the above serve as an illustration of the difference and the relationship between them for the purpose of this chapter. The reason it is important to note the difference is to be able to comprehend digital technology in a more nuanced way. Whether a digital product deals with data, information, knowledge, or intelligence, creates very different design parameters, hardware requirements, and fits into very different contexts with very different use cases.

Products that deal with data, information, and knowledge are the most common types of digital products and services that you have at your fingertips (i.e., laptop, tablet, or mobile phone). Think of the Word or Pages software, or your browser, social media feed, or your video player app. Some applications deal directly with data, such as Excel or a specialized data analysis software. Some applications may even go so far as to dabble with the concept of intelligence, such as predictive software and the early precursors to artificial intelligence (such as machine learning and complex adaptive algorithms) that allow the user to outsource a part of the decision making process. The most common current example of this type of technology is your favorite search engine.

Artificial Intelligence

We have come a long way in the field of artificial intelligence (AI), though there is still boundless debate on what exactly it is, what it is not, and what we can use it for, and how (and whether) it can improve some part of our reality. The debate is in itself fascinating as what we are in essence exploring is our comprehension and conceptualization of consciousness itself. What we have realized through attempting to re-create consciousness through digital technology (e.g., bots) is that our own consciousness is based on sets of algorithms that we can alter and update at will (if we have developed within ourselves advanced enough self-reflection and self-reflexivity skills to do so). We went from being a "black box" that was largely a mystery in how it operated to a biocomputer with clearly identifiable and observable (and more importantly constantly evolving and *changeable*) set of algorithms that define who and what we are and what we do. We also began to realize that there is intelligence everywhere, and humans are in a lot of ways at the lower end of the intelligence spectrum (in comparison to some natural phenomena or other biocomputers, such as fungi and dolphins).

The most important part of the debate is the role that artificial intelligence (not simply machine learning, but the capacity for something to make creative/unpredictable decisions on its own) plays in our future. Do we follow suit of dystopian or utopian fantasies and have it take over our world? Unlikely. When we look at the evolution of humans and technology, it has always been a symbiotic relationship (i.e., coevolution). Technology extends our capacity, rather than dominates us. The concept of "domination" is at its core a low-level intelligence concept. Any AI advanced enough would unlikely devolve into our early mistakes as a species. What we can extrapolate from all we know of intelligence is that the higher the intelligence, the more peaceful, kind, empathetic, and beneficial. The reason is simple: an intelligence advanced enough sees the wholism that is our reality, and the focus then shifts to harmony (e.g., equilibrium) and optimization rather than destruction or attempts to improve one part at the detriment of another.

The increase in academic and industry interest in AI in the 21st century has been fueled by the realization that pace of technological progress is exponential and not linear. Exponential progress has been mapped in advancements in hardware and software, including computational power over the last 60 years. Anyone can feel the changes increase quicker and quicker even in the last decade. Just look at what we can do now with a mobile device vs. a desktop machine in the 1990s. What this means is that technology will advance far enough to make AI (and other more advanced technology) a core part of our being in this lifetime. How might that look like in practice? Well, that is a question to those creating and adopting the tech. What will happen to our identity and sense of self is open to debate.

Beyond Legacy Systems

Luckily, we are more prepared to deal and adapt to major leaps and advancements in technology than ever before. The reason that information technology has arguably been the most significant evolutionary step of the 20th century is that it allowed us to record, explore, and compare paradigms. Tracing from instantiations (words, images, sounds, video, interactive experiences, etc.) to the axioms (and other underlying ontological and epistemological assumptions) and mechanisms that constitute perspectives. What we learned is that "who are we" is a flexible and constantly changing concept and that we are completely responsible for ourselves and our reality (not the concept of "god" or an institution or simply "someone else"). Arguably, for the first time in history, we have the complete freedom to think of ourselves as we see fit and change that at will (even if we cannot share that version of us with others due to the legacy system constraints of our material reality, though that is rapidly changing). Through digital media, we gained a sense of freedom and (relatively) consequence-free exploration of who we are and who we can be.

The point of profound empowerment and liberation of the self came when we first realized that the legacy systems[2] we inherited were almost all based on the core concept of "enslavement." Putting together information from across the world illuminated that "enslavement" was the dominant mechanism for tribal organizational systems that evolved into what we deem to be modern society. The global history of colonization, exploitation, slavery, oppression, punishment, control, and attempts to eradicate difference has created a mental cage from which we are only now beginning to emerge from and starting to create far more mature and sustainable alternatives. Both the wealthy and the poor have always been in the same boat, just with different cages.

The concept of "enslavement" was so prolific that it was largely unnoticed and passed down generation to generation through cultural inheritance. In some cultures, "enslavement" was a core concept around which identity was formed at all levels (e.g., in feudal systems). The 20th century was the first in history where this core concept was brought to light on a global scale, and we began to design with other much more wholesome and sustainable concepts at the core. Concepts such as sustainability, equality, kindness, and empathy. We saw a new renaissance in thought, where quality of life for all of life became important. Prominent theorists of the time may argue that this shift was due to the realization of our inevitable extinction due to the debt of legacy systems that we were taught to perpetuate; however, regardless of the reason, the shift has happened. It is important to note that, in general terms, some parts of the world (namely those saturated with digital technology) can be observed to be ahead of the curve than those without it.

---

[2]The term "legacy system," although originally pertaining to technology, in this book refers to any system of thought or practice.

What effect has this had on identity? A profound one. For the first time we have been able to design our own identity and base it on any ontological and epistemological assumptions that we want. With any axioms and core concepts that we choose. This is the first time the human species has had a chance at true liberation. The liberation of course comes from advancement in our own intelligence (at an individual and collective level). The advancements are fueled (if not directly caused) by the information and knowledge that we are able to collect and refine as a species. The exponential nature of this phenomenon gives great hope, as we have begun to realize that we do not have much longer left as a species unless we consciously change every aspect of our modern life. The planet simply cannot support our habits as they are for much longer.

How has that affected our digital products and services? We changed not only *what* but more importantly *how* and *why* we create. When digital technology first emerged, it was largely made by and for experts in technical fields. That made it extremely difficult for an average human to engage with it. Think of early DOS computers vs. your current laptop. As technology became more mainstream, more uses were found for it, more R&D was funded, and technology creators became more concerned with accessibility and ease of use rather than the more technical concerns of prior decades. Interestingly, in the 21st century, the core focus of technology creation has shifted to *design* of technology.

Design of technology has really become a common priority for almost every industry in the 21st century, with the focus of every development project being the design and alignment of design and user wants and needs. Before that it was still a largely specialized technical field where end users were not top priority. In the early days of software development, users were often forced to adapt to technology, as the tech was not designed with the experience of users front of mind. Professions such as User Experience Design, Service Design, and Digital Design emerged in response to the market demand for accessible tech. Eventually, core components of digital technology (at the back end) evolved enough for anyone to be able to create software without the massive investment seen in prior years and the focus shifted to the interface and user experience for the competitive advantage. In 2020, anyone can build a website or a mobile app by using digital autogeneration service providers.

How has this shift affected digital products and services? It made it better and more seamlessly integrated into our daily habits. Digital now matches us, rather than the other way around. How has this come to be? Digital has always borrowed from all types of industries. The most recent phenomenon is rapid proliferation of such approaches as Design Thinking to create digital products and services. In such approaches, every decision is based on researching and deeply comprehending users and their lives through immersion, empathy, and the intent to genuinely solve the problems of those who will use our products and services. It's the epitome of selfish altruism: design for what matters to users, and the product or service is guaranteed to be a success.

As consumers (or users), our core responsibility is to educate ourselves, to become discerning (so that the types of products and services created through legacy systems do not find their way into our lives), and to genuinely and honestly help those that truly care about improving our lives (as we want it) to design better products and services for us, with us, and to the benefit of both us and them.

Luckily, this is extremely easy to do because of the nature of open markets, information available, and the plethora of products and services that has flooded us with choice.

Where We Are and Where We Are Going

As the film *The Matrix* (1999) popularized, we were all born into a set of legacy systems that we were blind to from birth because we were taught not to question what was deemed "sacred" or "untouchable" in a culture. Cultures varied and not the same things were uniformly considered "sacred." Some cultures held life sacred, others wealth. Some free will, some control. Some unity with nature, some exploitation of everything in reach. With the age of digital media, we (as a species) were finally able to compare and contrast different cultures and the paradigms that constituted them and, with the capacity to compare, were able to see everything for what it was. Then we were able to take the best, leave the rest, integrate, upgrade, refine, and optimize. That is how we got to the "melting pot of culture" that is our digital reality.

In the last 40 years, we have given ourselves the gift of sharing our personal (subjective) experiences on mass and allowed for enough information to be circulated for us to develop truly novel ways of seeing ourselves and our place in the continuum of life. New paradigms emerged that retreated from the dominant narratives of legacy cultures and allowed for the emergence of both a global uniform digitally enabled culture and pockets of subcultures based on compatible ontologies/epistemologies. We have learned to see behind the facade of everything and to trust ourselves more than whatever "authority" we were taught to fear or obey (all of which we realized was at some point in time self-appointed and, in the grand scheme of things, powerless and meaningless) and learned to listen to true experts—those that contributed to shedding light on the mechanisms and ways of being and extending those (thereby improving them) through personal experimentation (i.e., on themselves, not on others) and those that have paved the way for different ways of being and doing by standing up to the dominant norms of their time.

So, now that we reached a certain level of awareness and realized that the ways of the past are not the ways of the future, now what? Where are we going? We have two options: extinction or enlightenment. Does that mean that we will need to change almost everything about our life? Yes. Is that an opportunity to redesign everything? Yes. Is that the ultimate place of empowerment? Yes.

This is the reality that the digital designer needs to consider in order to create products and services that are of value and that deeply resonate with and improve quality of life. The priorities of users are shifting from the pipe dreams of early 20th-century advertising to something that provides true meaning and value to them. After all, what good is a pursuit of wealth if there is no longer a reality in which it has any value?

It is important to note that any perspective and prediction of where we are going is largely shaped by what we consider to be of value. It is also important to note that value is not static and is rather elusive because it is relational (i.e., something can only be considered as "valuable" in relation to something/ everything else). Something is of value to someone or something when it has meaning and purpose, and improves something (similar to the difference between invention for the sake of invention and deliberate innovation). When context changes, what is considered of value changes.

Redefining "value" in relation to the individual and the collective is part of the transformation of our identity and our roles. Exploring and consciously redefining value is arguably the only way that we can purposefully, meaningfully, and strategically design for change. When we value sustainability, profit through exploitation becomes devalued. Why? Because exploitation is always unsustainable (even if it can last a certain amount of time, it always breaks down as a mechanism). Another reason is that by exploring and comprehending true value (i.e., true to the individual), we can begin to map our current state in detail and see opportunities for change and for creating and/or redirecting value. For example, it is difficult to temp someone with an unattainable ideal and keep them running after it like a donkey chasing a carrot on a stick (the core mechanism for most early 20th-century advertising and sales) if they see no value in that ideal. When someone forms their own sense of value and purpose derived from the intangible ideals that they embody and aspire to, then the only way to get them to make the choice to engage is to appeal to their core, not to something arbitrary.

Value is the barometer, if not the compass, for change.

A good exercise to start to figure out what you (and your users) value: think of what you truly want[3] right now. Imagine you have unlimited quantities of it, and ask yourself (or your users) "Now what?". Repeat at least 9 times (or until the point where the core axiom of value is reached). Record as you go and map patterns. This simple exercise helps raise thinking to beyond the immediate wants (that, at first, are often pre-programmed and untrue to the core drivers of the individual). Alignment through core values is always stronger than alignment through any other means.

And if we dig deep enough, we are likely to reach the same realization that philosophers, authors, poets, and artists have alluded to over the centuries: what we all *truly* want and value is really the same and it is simple. And you cannot buy it in a pill or a bottle or a box. What has distracted us from what we truly want are the remnants of our evolutionary passage that are now falling away, like a snake shedding its skin.

Identity, Paradigms, and Digital

There are infinite possibilities for a new identity in the 21st century. As explored in Chapter 2, the way that someone constructs their identity shapes their entire being: they sense of self, their actions, their choices, and their role in their community. Identity can be constructed through static attributes or dynamically procedurally generated from a set of core values and algorithms (i.e., values ➤ principles ➤ ethics ➤ behaviors). Some transitional identities may be a combination of the two types.

Static attribute identities were the most common type in the majority of legacy cultures and social systems. These types of identities were based on arbitrary attribution of meaning to something, such as gender, age, class, skin color, locale, and more. These types of identities are very restrictive as they do not allow for the evolution of an individual over time, and therefore go against the very nature of life, which is to change and evolve. Digital technology has fundamentally disrupted these types of identities as it allowed for individuals to explore identities based on something deeper and richer than an arbitrary attribute and its culturally assigned meaning. This was the first step to the identity fluidity most digital migrants (i.e., those born before the digital era who adopted the technologies) and digital natives (i.e., those born into the digital era) experience as their norm.

---

[3]A note on "wants" and "needs": from experience designing across cultures, these two words are very loaded and should be used with caution. In some cultures the word "want" is honored as "empowered" while "need" is condemned as "weak." In other cultures "want" is an unattainable ideal (and often deemed "frivolous"), while "need" is the practical reality (and itself considered a "virtue"). Watch for cross-cultural paradigm clashes of those your converse with for truly meaningful connections and impactful insights.

How did technology enable this shift from static to procedurally generated (and therefore fluid) identities? By allowing for exploration of the deeper drivers of human beings: their core values. Think of the computer game series *The Sims*. The core concept of the game is to place the user in a certain "god mode" where they are able to control the actions and life circumstances of a set of humans they view in third-person perspective on the screen. The game can be played in what feels like an infinite variety of ways, though it closely mimics the life passage of the average human being in the early 21st century: get a job, to get money, to get things, then get a better job, to get more money, to get more things, and so on. The game does allow for cheat codes that give much more freedom to the user to explore the possibilities of constructing and decorating homes, creating characters and their families, exploring different social dynamics, and so on. Each character also has a set of metrics and criteria for skills, happiness, hunger, fun, hygiene, and so on.

The game in a very literal sense allows for the user to explore different ways of living and being and explore all kinds of realities that either closely mimic their own or allow for them to explore other ways of being by vicariously living through their created characters. This may seem no different to any other game, whether a car racing game, a first-person shooter, or a medieval-style battle game. However, *The Sims* goes far beyond entertainment and allows the user to dive deeply into their darkest fantasies, dreams, and shortcomings by providing a mirror to their consciousness and sense of self. It allows for what matters to them and their true desires to be explored in a safe context (e.g., killing characters in *The Sims* in elaborate ways is fundamentally different from killing anything in real life, or simply having the experience of being a different gender or observing the effects of neglect on their character). The virtual context provides a safe way for someone to explore themselves without consequence (in their external reality) and for them to reach the "Aha!" moment of an epiphany when their skills of self-reflection and reflexivity become advanced enough. The person playing the game and realizing that they spent three months making the lives of their characters lovely while doing absolutely nothing to improve their own, in that realization, experiences a profound moment of self-reflexivity that has a core impact and shapes the rest of their lives. Similar phenomena occur through shaping the user's cognitive system on how they might take care of themselves later in life, how they might want to design their house, what kind of family they may want to have, what to focus on, and what to avoid.

This type of exploration enables identity to retreat from static attributes and their assigned cultural meaning, and allows for identity to form around actions, and the mechanisms and core values that drive those actions. *The Sims* is one example of a plethora of games, forums, social media platforms, and other digital technologies that pivot someone to a place where they are forced to confront themselves and their choices. What all these technologies have in common is their capacity to allow someone to be more and something other than what they are in their material reality. The choices that become available and the exploration that follows are the true catalyst for personal discovery, realization, and transformation.

Once one undertakes the challenge of knowing themselves (beyond what they see reflected back at them in their Facebook profile page, which as discussed before is the equivalent of a showroom), there is no going back to simpler comprehension of who they are and what truly matters to them. To "know thyself," as popularized by the Oracle of Delphi legend, is the ultimate position of clarity and empowerment. Because without it, we cannot know where we are, and without knowing where we are, we have no conscious choice of where we go next. Luckily, we now live in an age where the opportunity of self-discovery is available in abundance.

How might we as consumers (1) map our current state, (2) decide where to go, and later (3) transform into what we want to become (rather than where we came from)? Through the instantiations that we create. Words, images, texts, all the things that we have recorded in digital products and services. These types of products allow for us to collect incredible amounts of information on ourselves and our behaviors. There are digital products that allow us to track our mood, journal our thoughts, capture moments we want to remember, track our fitness and dietary habits, monitor sleeping patterns, help us reduce addictive tendencies, and upskill us in any way we want. All that information provides a much more complete picture of who and what we are and where we are going. It provides us a powerful set of data and information about us, our values, and our choices. What we have is a fragmented set of analytics[4] for identity.

The most important part of the entire phenomenon of digital transformation: the capacity for change of core paradigms that we operate through. The only change that matters is the change in our beliefs and how we see ourselves and the world. Change beliefs, and the whole world transforms. History has demonstrated that in abundance. Progress can only be truly felt through paradigm change.

---

[4]On who owns data: the data that you generate through an app as a user is arguably yours, not who owns the server. Be careful in choosing the digital products that you use, as most early 21st century digital products treated your data as theirs and used it to generate incredible amounts of profit that users did not receive a share of (and sometimes that profit was generated at the expense of users and their basic rights). Remember: in digital, if you are not paying for the product, you are the product.

Implications for Digital Designers

The pace of progress is exponential. What this means is that our paradigms are changing exponentially as well. This has a variety of implications for digital designers. When we design something that is a digital product and/or service, what we are designing first and foremost is an extension of our users, and also an experience that forms their habits (and their habits shape their life and their future). That is a lot of responsibility and this responsibility goes beyond designing interfaces that look and feel good to use. The responsibility extends to the very nature of the technology and the effects it has on the users and all of life.

When starting the journey of creating a new digital product and/or service, start with getting to know your intended end users and their lives, wants, needs, desires, and aspirations. Look for how they construct their identity and how their perception and experience or reality is shaped by the paradigms they adopt and operate through. Help them explore their say-do gap, and why it exists. Ask them what makes them happy and what makes them sad. Empathize with their journey and be kind to those who are not like you.

This research forms the foundation for creating products and/or services that truly resonate, add value, and help the users go in the direction that they want to go. Design with the end goal of improving quality of life for all in mind, and do not compromise on what matters to your users and what contributes to creating a better reality for all. Our future is in your hands.

Put simply:

- Design for good.
- Design for change.
- Design for empowerment.
- Design for liberation.
- Design for life.

Conclusion

Empowered Digital Designers

Digital is one of the most important aspects of our lives in the 21st century. It has become an inseparable extension of our selves and given us the opportunity, for the first time in history, to limitlessly explore who and what we are, and who and what we want to become. We have been able to connect on levels previously unimaginable, completely unencumbered by limitations of locale. For this we have to thank the creators of digital.

In order to help us orient to where we are and where we have come from, this book provided an overview of the emergence of digital, its changing role, and the effects and implications of digital on the realization of the potential of our future. We, as digital designers, play a crucial role in helping our species shift from legacy constraints of the past, and to shape the formation of our future. Our practice has evolved with the needs of the collective, and it is my sincere hope that the perspectives outlined in this book are able to assist the journey.

The following is a brief summary of the topics explored in each chapter.

In **Chapter 1**, "**Digital Evolution**," we looked at how digital technology has come to be such a prominent feature in modern life. The chapter looks in detail at the process of the coevolution of humans and technology, using the meta-theoretical framework of Universal Darwinism in order to view the four core passages of information evolution: biology, technology, culture, and cognition. The chapter provides digital designers with a perspective on digital

© Anastasia Utesheva 2020
A. Utesheva, *Designing Products for Evolving Digital Users*,
https://doi.org/10.1007/978-1-4842-6379-2_7

that places it in context of human evolution, and evolution of information more broadly. The core question that the chapter aims to address: *how might we design for purposeful and meaningful change and, most importantly, assume our full responsibility in creating and driving these changes?*

Chapter 2, "**Evolution of Identity**," further addresses this question, by focusing on the core driver of human decision making: their identity. The chapter provides a conceptualization of identity suitable for digital designers and explores how human identity has been extended and transformed through technology (specifically digital). We explore the important shift in identity from static to procedurally generated identities and the impact this shift has had on users. This chapter also looks at the evolution of identity and explores the differences between assigned and enacted identities. Ultimately, the chapter provides a foundation for strategically using identity in digital design and outlines implications of the formation of digital identity on local and global social dynamics.

Chapter 3, "**Designing for Evolving Users**," further extends the discussion of identity in digital design by providing a perspective on how to design digital products and/or services for evolving users. The chapter begins to unravel what it means to be a digital designer in the 21st century and the responsibilities that digital designers face when creating products and services. We look at digital design as designing intangible "experiences" rather than tangible, static "objects." We also explore how the role of users changed from "passive" to "active" and the implications of this shift on the practice of digital design. The role of digital designers is argued to extend the reach of the user such that they have complete control of the experience and its outcomes and that the core of what we are actually extending in digital design is the "mind." The chapter also provides a summary of core digital design principles and processes.

In **Chapter 4**, "**Design Thinking in Action**," we look at a specific type of digital design approach: Design Thinking. The chapter provides a brief overview of Design Thinking and a summary of how the approach can be applied to digital design. The value of Design Thinking is derived from the rapid evolution of ideas and evidence-based design decisions, which are argued as vital for successful design in this field. The chapter also provides a high-level summary of a Design Thinking toolkit and how to apply this toolkit to design higher-quality products and services. The chapter aims to provide digital designers just starting out with practical guide on how to design for real needs of real humans.

Chapter 5, "**Design for Change**," looks at how we might strategically design for change. The chapter provides an overview of what it means to strategically design digital and revisits the role of digital in modern life. We explore how we might design to improve quality of life through the products and/ services that we create. Specifically, we look at how we might evolve identity through digital. The chapter also provides a simple exercise on how to map your own identity: a self-reflective tool vital for all designers. We also explore how we can monitor and design for change through paradigms and paradigm shifts over time.

Finally, **Chapter 6**, **"Digital Trends,"** looks at the significant digital trends over the past century. The chapter focuses specifically on automation, information communication technology, and artificial intelligence. The chapter also provides a high-level discussion of the impact of exponential digital progress on life. The important focal point is that, as digital designers, we changed not only *what* but *how* and *why* we create digital products and/or services. The chapter highlights that we, as creators of digital, have also shifted from focus on technical aspects of technology creation to *design of* technology. The chapter also looks at the shifting concept of value and argues that for successful design we must aim to align through core values rather than through any other means. The chapter concludes with the exploration of the intersection of identity, paradigms, and digital. We see the value of digital transformation in the capacity to change the core paradigms that we operate through and suggest that progress can only be truly felt through paradigm change.

Each chapter provides an exploration of a component of digital design, and the intent is to help shift perspectives to allow for better and more strategic digital design to emerge. These discussions are by no means extensive but serve as a basis for future exploration, discussion, and evolution of the practice.

Thank you for coming on the journey of digital design. Our future is better for it.

Index

A

Agile approaches, 55

Artificial intelligence (AI), 94

Assigned identity, 32

Automation, 12, 92

Automobiles, 12

B

Biological evolution, 4, 9
  adaptation, 11
  design knowledge bases, 12
  genetic code, 29
  genetics, 10
  human, 10
  modern civilization, 11
  multiple species, 10
  theories, 7

Biological technology, 12

C

Change perception, 77

Cognition, 4, 26

Cognitive evolution
  accommodation, 28
  accounts, 26
  assimilation, 28
  assumption, 27
  complexity, 27
  core mechanism, 27
  equilibrium, 28

major systems, 27
memetic skills, 27
mimetic replicator entities, 27
outside-inside principle, 26
paradigms, 28
representational systems, 27
schema formation, 28

Collective identities, 78, 81

Complex/elegant technology, 12

Copy-the-instructions, 24

Copy-the-product, 24

Cultural evolution
  cognitive models, 23
  complex meaning systems, 22
  continuous accumulation/refinement, 20
  elements, 20
  externalize/internalize, 20
  fake information field, 24
  historicity, 20
  labels/categorizations, 20
  material instantiations, 24
  mechanisms, 23
  medium-specific language, 21
  memeplex, 21
  memes vs. genes, 22
  memory, 22
  paradigms, 22
  philosophical paradigms/axioms, 20
  prevalence/rootedness, 20
  replicator entities, 25
  representational content, 24
  representations, 22, 23

© Anastasia Utesheva 2020
A. Utesheva, *Designing Products for Evolving Digital Users*,
https://doi.org/10.1007/978-1-4842-6379-2

GPSR Compliance
The European Union's (EU) General Product Safety Regulation (GPSR) is a set
of rules that requires consumer products to be safe and our obligations to
ensure this.

If you have any concerns about our products, you can contact us on

ProductSafety@springernature.com

In case Publisher is established outside the EU, the EU authorized
representative is:

Springer Nature Customer Service Center GmbH
Europaplatz 3
69115 Heidelberg, Germany

www.ingramcontent.com/pod-product-compliance
Lightning Source LLC
Chambersburg PA
CBHW071220050326
40689CB00011B/2389

* 9 7 8 1 4 8 4 2 6 3 7 8 5 *